Future Energy

Future Energy

Editor

Shewta Saini

Future Energy
Edited by **Shewta Saini**

Printed in 2017

ISBN: 978-1-68117-381-8

Library of Congress Control Number: 2015941567

© 2016 by

SCITUS Academics LLC,
616, Corporate Way, Suite 2, 4766,
Valley Cottage, NY 10989

www.scitusacademics.com

Contents

Preface

Future Energy is a former accreditation scheme for green electricity in the United Kingdom, designed to support and stimulate electricity generation from renewable energy sources. The coming decades will bring big changes in energy systems throughout the world. The systems are expected to change from central power plants producing electricity and maybe heat for the customers to a combination of central units and a variety of distributed units such as renewable energy technologies or fuel cells. Future Energy will allow us to make reasonable, logical and correct decisions on our future energy as a result of two of the most serious problems that the civilized world has to face; the looming shortage of oil (which supplies most of our transport fuel) and the alarming rise in atmospheric carbon dioxide (resulting from the burning of oil, gas and coal and the loss of forests) that threatens to change the world's climate through global warming. The Future of Energy discusses the sources, technologies, and tradeoffs involved in meeting the worlds energy needs. A historical, scientific, and technical background set the stage for discussions on a wide range of energy sources, including conventional fossil fuels like oil, gas, and coal, as well as emerging renewable sources like solar, wind, geothermal, and biofuels.

Editor

Chapter 1

Emerging Electrochemical Energy Conversion and Storage Technologies

Sukhvinder P. S. Badwal, Sarbjit S. Giddey, Christopher Munnings, Anand I. Bhatt and Anthony F. Hollenkamp

Commonwealth Scientific and Industrial Research Organisation (CSIRO), Energy Flagship, Clayton South, VIC, Australia

ABSTRACT

Electrochemical cells and systems play a key role in a wide range of industry sectors. These devices are critical enabling technologies for renewable energy; energy management, conservation, and storage; pollution control/monitoring; and greenhouse gas reduction. A large number of electrochemical energy technologies have been developed in the past. These systems continue to be optimized in terms of cost, life time, and performance, leading to their continued expansion into existing and emerging market sectors. The more established technologies such as deep-cycle batteries and sensors are being joined by emerging technologies such as fuel cells, large format lithium-ion batteries, electrochemical reactors; ion transport membranes and supercapacitors. This growing demand (multi billion dollars) for electrochemical energy systems along with the increasing maturity of a number of technologies is having a significant effect on the global research and development effort which is increasing in both in size and depth. A number of new technologies, which will have substantial impact on the environment and the way we produce and utilize energy, are under development. This paper presents an overview of several emerging electrochemical energy technologies along with a discussion some of the key technical challenges.

INTRODUCTION

In view of the projected global energy demand and increasing levels of greenhouse gases and pollutants (NO_x, SO_x, fine particulates), there is a well-established need for new energy technologies which provide clean and environmentally friendly solutions to meet end user requirements. It has been clear for decades that renewable energy sources such as wind and solar would play some role in the modern grid with predictions varying on the levels of penetration and the effect that these renewable power sources would have on the stability of national grids. The role that renewable energy will play in the future energy mix is now becoming more obvious as this sector matures. As higher levels of renewable energy are integrated into national grids a greater understanding of the effect of their intermittent nature is becoming wide spread. This can result in significant mismatch between supply and demand. In addition to the changes to the power generation infrastructure, the integration of smart meters is leading to a market where energy use can be easily measured in real time. In order to maximize profit, privatized power generators and grid suppliers are increasingly promoting the use of strong financial incentives to be levied on power users to change their electrical energy usage habits. This has led to a defined cost being associated with the previously largely invisible tasks associated with managing power generation and large distribution grids. This clear cost signal has led to increased demand for energy storage for load-leveling, peak load shaving, and providing power when the renewable energy is not available at almost every level of the power generation market from small scale domestic devices to large scale grid connected systems. In the future energy mix, electrochemical energy systems will play a key role in energy sustainability; energy conversion, conservation and storage; pollution control/monitoring; and greenhouse gas reduction. In general such systems offer high efficiencies, are modular in construction, and produce low chemical and noise pollution.

In real-life applications, the limitations of single power generation or storage technology based energy solutions are now being recognized. In many instances the requirements (e.g., response time, power capability, energy density, etc.) for energy storage technologies far exceed the performance limits of current energy technology solutions and in some instances also exceed the theoretical limits of a given technology. Thus, there is a substantial current and future (new applications) global demand for hybrid energy solutions or power sources to optimize cost, efficiency, reliability, and lifetime whilst meeting the performance requirements of the applications. In this regard many electrochemical energy technologies are expected to play a key role.

In most electrochemical energy technologies, the electrode and electrolyte materials must possess the required ionic and electronic transport properties and a great deal of research is still to be performed at a fundamental level to study and optimize the electrochemistry of candidate materials, composites, and assemblies (such as catalyst and interface designs). Practical materials must operate in a multidimensional space where optimum electrochemical properties must co-exist with secondary properties such as chemical stability, compatibility with other components (thermal expansion co-efficient, strength, toughness, etc.) and at the same time they must be amenable to be fabricated into the required shapes and forms at acceptable cost. Materials and properties need to be carefully tailored and matched to suit a technological application and the environments in which they are to be used. At higher operating temperature, these requirements are more stringent and, in fact, they become critical at temperatures above 500°C. At these temperatures, other issues, such as gas sealing, interface compatibility and stability, and the design of support structures and containment materials are as challenging to solve as the technical issue directly associated with the electrochemical cells. Many materials and system integration complexities exist and these are being resolved through investments in experimental developments and through theoretical modeling. Once these challenges are solved, the practical applications of electrochemical energy technologies are numerous.

Some of the electrochemical energy technologies developed and commercialized in the past include chemical sensors for human and asset safety, energy efficiency, industrial process/quality control, and pollution control/monitoring; various types of fuel cells as clean energy devices for transport, stationary and portable power; a range of energy storage batteries; electrochemical reactors for fuel and chemical production; electrochromic smart windows for optical modulation and building efficiency; ion transport membranes for air separation; and supercapacitors (Guth et al., 2009;Scrosati et al., 2011; Yang et al., 2011; IPHE, 2012; Sbar et al., 2012; Wilson et al., 2012; Akhil et al., 2013; Carter and Wing, 2013; Harrop et al., 2014; Stiegel et al., 2014). While these technologies continue to be optimized for cost, lifetime, and performance, there is a substantial growing demand (multi billion dollars) for advanced electrochemical energy systems such as high energy density batteries for transport vehicles and stationary energy storage; next generation fuel cells with high efficiency, better performance, and long life; membrane reactors for value added chemical production; gas separation devices in medical and power generation; and hybrid fossil fuel/storage/renewable energy systems. In this paper an overview of some more recent and emerging electrochemical technologies is given and some of the fundamental challenges facing technology development are discussed.

HYDROGEN PRODUCTION TECHNOLOGIES

Hydrogen is considered to be an important energy carrier and storage media for a future hydrogen economy. Hydrogen offers a sustainable energy future for both transport and stationary applications with near zero greenhouse gas emissions especially when generated by splitting water and combining with renewable energy sources (solar, wind, ocean). Since most renewable energy sources are intermittent in nature, hydrogen can act as a storage media for load leveling and peak load shaving. It can be generated when abundant renewable energy is available and stored and converted to power and heat in a fuel cell or combustion engine as per load demand based on end-use applications. A number of different electrochemical technologies are under development and these will be briefly reviewed in the following sections.

LOW TEMPERATURE WATER ELECTROLYSIS

Hydrogen can be generated by electrolyzing water at low temperatures (LTs) (<100°C) or electrolyzing steam at high temperatures (HTs) (>700–800°C). The LT electrolysis systems employ either an alkaline (hydroxyl ion conducting) solution as the electrolyte or a polymer membrane (proton conducting) as the electrolyte (Figure 1) (Ursua et al., 2012; Badwal et al., 2013). The hydrogen generation by utilizing a LT electrolyzer compared to that produced by natural gas (NG) reforming or coal gasification, offers a number of advantages such as on-site, on-demand (distributed) generation, high purity hydrogen, and unit modularity. Furthermore, such systems offer fast start-up and shutdown, and good load following capability that makes them suitable for integrating with intermittent renewable energy sources such as solar PV and wind generators. In LT systems, polymer electrolyte membrane (PEM)-based systems offer additional advantages over alkaline systems such as higher current densities (small foot print in terms of kgs per hour hydrogen generation capacity per unit stack volume), all solid state system requiring no alkaline solutions or electrolyte top-up, and higher purity hydrogen and hydrogen generation at significantly higher pressures (Badwal et al., 2013).

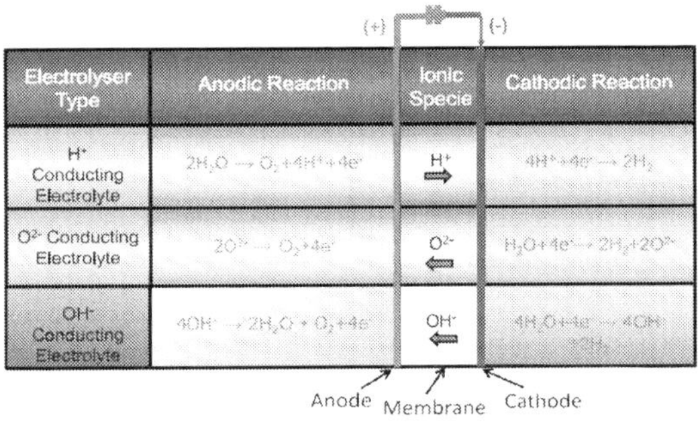

Electrolyser Type	Anodic Reaction	Ionic Specie	Cathodic Reaction
H$^+$ Conducting Electrolyte	$2H_2O \rightarrow O_2 + 4H^+ + 4e^-$	H$^+$ $\Rightarrow$	$4H^+ + 4e^- \rightarrow 2H_2$
O^{2-} Conducting Electrolyte	$2O^{2-} \rightarrow O_2 + 4e^-$	O^{2-} $\Leftarrow$	$H_2O + 4e^- \rightarrow 2H_2 + 2O^{2-}$
OH$^-$ Conducting Electrolyte	$4OH^- \rightarrow 2H_2O + O_2 + 4e^-$	OH$^-$ $\Leftarrow$	$4H_2O + 4e^- \rightarrow 4OH^- + H_2$

Anode Membrane Cathode

Figure 1: Operating principles of low and high temperature water electrolysis with different electrolytes.

A typical electrolyzer system may comprise of the electrolyzer stack and balance of plant (BOP) subsystems for water deionization and circulation to anode chamber, water/gas separation (for oxygen and hydrogen), heat management, hydrogen drying and storage, and a DC power source. The stack constitutes a number of cells or membrane electrode assemblies (MEAs), assembled between bipolar metallic interconnects. The interconnects supply and collect respectively the reactants and products from cells and connects the cells in series. Further details on MEAs and electrolyzer stack assembly can be found in references (Clarke et al., 2009; Giddey et al., 2010; Ursua et al., 2012).

A number of companies (Proton OnSite, Giner Electrochemical Systems, Hydrogenics, Horizon, ITM Power) are now selling LT electrolysis systems at prices which are commercially not competitive with other processes for hydrogen production (e.g., NG steam reforming). Thus, a number of challenges related to high cost of commercial units, lifetime and net efficiency still remain.

Furthermore, the hydrogen generation by electrolysis is an energy intensive process and most commercial electrolyzers require an electric power input of 6.7–7.3 kWh/Nm³ (~50–55% efficiency based on HHV) of hydrogen (Badwal et al., 2013), and this increases the cost of hydrogen production and advantages of hydrogen as a clean fuel are lost if the electricity is supplied from fossil fuel resources. However, if the electric energy input can be supplied from renewable sources of energy and the electrolyzer system efficiency increased to 75–80%, then the technology becomes more attractive. The LT electrolyzers can easily operate with a large load variation and thus are highly suitable for integration with intermittent renewable energy sources. Figure 2 shows a concept of a renewable energy system based on hydrogen generation by direct coupling of an electrolyzer to solar PV and a wind generator. This type of system can be used to store hydrogen and operate a PEM fuel cell to provide power at times when renewable energy cannot meet the load demand. The other components shown in the diagram are a diesel generator as a backup, and a hot water storage tank to collect hot water from the PEM fuel cell that can be used for daily needs of a house.

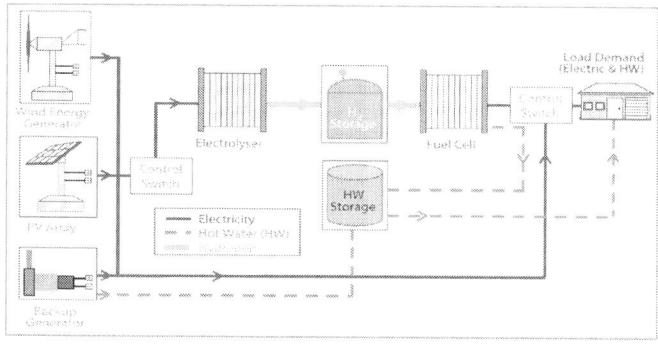

Figure 2: Overall concept of a hydrogen renewable energy system for distributed power generation.

The direct coupling of an electrolyzer to renewable sources of energy must ensure that there is a maximum transfer of electric energy from the renewable source to the electrolyzer to produce hydrogen. By incorporating appropriate maximum power point trackers (MPPT) and DC-DC converters to meet these requirements, a number of systems have been demonstrated in the past. However, this substantially adds to the cost and makes the renewable energy—hydrogen generation system economically less viable. Therefore, it would be beneficial if the renewable source of energy is directly coupled to the electrolyzer without any electronics or control system, and also without losing on the energy transfer to the electrolyzer. There have already been studies and demonstrations for hydrogen generation by coupling PEM-based electrolyzers to solar PV (Arriaga et al., 2007; Clarke et al., 2009) and to a wind generator (Harrison et al., 2009).

Figure 3 shows a typical example of matching the maximum power point (MPP) curve of solar PV array to the V-I characteristics of an electrolyzer (Clarke et al., 2009). The matching criteria are to achieve maximum transfer of energy from the solar PV system to the electrolyzer by matching the output of PV to the input power requirements of the electrolyzer. In the example in Figure 3, this was achieved by coupling 15 pairs of solar PV arrays in parallel to a 16 cell electrolyzer stack. The modeling of such a system showed that there will be on average 99.7% of the solar PV energy transfer to electrolyzer at all values of solar irradiance, and about 8% overall solar to hydrogen efficiency. Although the direct coupling of the renewable sources to an electrolyzer offers a relatively cheaper and more efficient way of generating hydrogen, there are two major challenges to this technology—first is on the relative sizing of the two units due to variability of the energy source (solar irradiance and wind speed) to achieve maximum benefits of coupling, and second is on the long-term performance of the electrolyzer on a continuously variable load. In a recent publication (García-Valverde et al., 2011), the authors have endeavored to tackle the first challenge by modeling polarization (V-I) curves of both, the solar PV and the electrolyzer. In relation to the second challenge, in a study carried out by NREL, a prototype electrolyzer was tested on a variable (wind generator) load profile for up to 7500 h with a small degradation in the electrolyzer performance (Harrison and Peters, 2013), however, the electrolyzer failed soon afterwards.

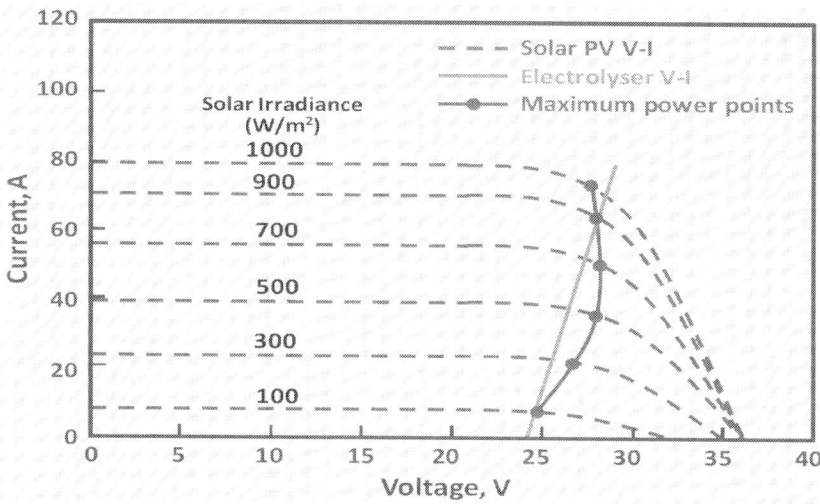

Figure 3: A typical example of matching maximum power point (MPP) curve of a suitably configured solar PV array to V-I characteristics of an electrolyzer. The example is for 15 pairs of solar PV arrays connected in parallel and a 16 cell electrolyzer. The data in the Figure has been taken from Clarke et al. (2009).

HIGH TEMPERATURE WATER ELECTROLYSIS

As discussed above, hydrogen can be readily produced via LT electrolysis at almost any scale using only water and electricity as the inputs. This process is well-established but requires a high input of electrical energy in order to produce the hydrogen. From a thermodynamic perspective at 25°C, 1 liter of hydrogen requires a minimum 3.55 kWh of electrical energy as an input. This increases to around 4.26 kWh when electrochemical cell losses are taken into account. If the electrolysis process is carried out at HT then it is possible to utilize some of the heat for the production of hydrogen. This contribution can be high with up to a 1/3rd of the energy required to produce the hydrogen coming from thermal energy at around 1000°C (Figure 4) (Edwards et al., 2002; Brisse et al., 2008; Laguna-Bercero, 2012; Ursua et al., 2012; Badwal et al., 2013). In Figure 4, the thermal energy input under cell operation may be slightly different due to internal heating of the cell resulting from current passage, however, due to the difficulty in making an estimate, it has been assumed to be the same as that under open circuit cell conditions. The HT electrolysis systems use an oxygen ion (O^{2-}) or proton conducting (H^+) ceramic as the electrolyte (Figure 1) (Edwards et al., 2002; Brisse et al., 2008; Laguna-Bercero, 2012; Ursua et al., 2012; Badwal et al., 2013). The process is the reverse to that of a solid oxide fuel cell (SOFC) with many similar materials used for cell construction. The thermal input required for HT systems can be supplied from different sources including renewable or sustainable energy sources or nuclear energy.

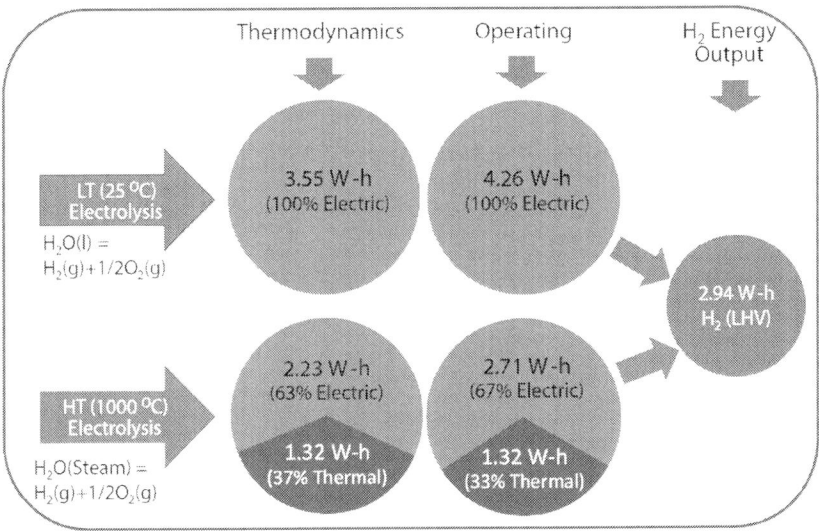

Figure 4: Break down of energy input for the production of hydrogen from electrolysis at 25°C and 1000°C. The data in the Figure has been taken fromBadwal et al. (2013).

A number of different systems have been proposed including the co-locating of the electrolyzer with a solar thermal source, nuclear power stations, or supplying heat produced from the burning of low grade fuels such as coal (Edwards et al., 2002; Fujiwara et al., 2008; Badwal et al., 2013). A number of systems and materials configurations have been trialed with zirconium-based oxide ion conducting electrolytes in conjunction with manganite-based anodes and metal cermet cathodes being the most commonly used materials (Ursua et al., 2012; Badwal et al., 2013). There have been a number of reasonably significant demonstrations of this technology (up to 15 kW) but no commercial or near commercial prototypes produced (Badwal et al., 2013). These trials have demonstrated the technical feasibility of this technology, however, cost, lifetime, and reliability remain as some of the key challenges (Badwal et al., 2013). If HT electrolysis is to be commercialized then there would need to be either a significant increase in the cost of hydrocarbon fuels or a significant reduction in the cost of HT electrolyzers. The HT systems, despite offering energy efficiency advantages due to thermal input, are still at early stages of development.

In order for hydrogen to be cost competitive with other hydrocarbon fuels, the US DOE have set a cost target of $3/kg of hydrogen. If electricity and water are the only inputs (as is the case at 25°C), this leads to the electricity cost needing to be well-below 0.06 c/kWh (Badwal et al., 2013). Although this is potentially feasible, the additional costs associated with compression, transportation, and distribution make the conversion of high grade electrical power from the grid directly to hydrogen uneconomical. However, if a suitable source of thermal energy can be used then electrical component contribution reduces significantly.

CARBON-ASSISTED HYDROGEN PRODUCTION

The use of hydrogen as a transport fuel in fuel cell or internal combustion engine vehicles is likely to increase due to the concerns over oil shortage and rising greenhouse gas and other pollutant emissions. Hydrogen is generated mainly from NG and coal involving three major steps requiring separate reactors, all operating at temperatures in excess of 500°C: (i) NG reforming or coal gasification to produce syngas (a mixture of hydrogen and carbon monoxide) at temperatures close to 800°C; (ii) water gas shift reaction to convert carbon monoxide to hydrogen and carbon dioxide at around 500°C; and (iii) H_2/CO_2 separation and gas cleaning. Hydrogen production by water or steam electrolysis in which the electricity is drawn from the grid is overall a highly inefficient process, in that it requires electric input of 4.2–5 kWh per Nm^3 and 6.7–7.3 kWh per Nm^3 of hydrogen for the electrolysis cell stack and system, respectively.

The participation of carbon in the anodic reaction of the electrolysis results in a drop in the thermo-neutral voltage from 1.48 to 0.45 V required for electrolysis of water near room temperature (Coughlin and Farooque, 1982), which can translate into reduction in electric energy input to 1/3rd compared to normal electrolysis. Thus, the remaining 2/3rd of the energy would be supplied from the chemical energy of carbon. The carbon-assisted electrolysis carried out at higher temperatures can result in further reduction in the required electric energy input due to increased thermal energy contribution into the process by lowering the thermo-neutral voltage further (Seehra and Bollineni, 2009; Ewan and Adeniyi, 2013). Figure 5 schematically shows the electrochemical reactions involved for carbon-assisted electrolysis carried out at temperature <100°C (LT) employing a proton conducting electrolyte membrane, and at HTs (>800°C) employing an oxygen ion conducting ceramic electrolyte such as yttria or scandia stabilized zirconia.

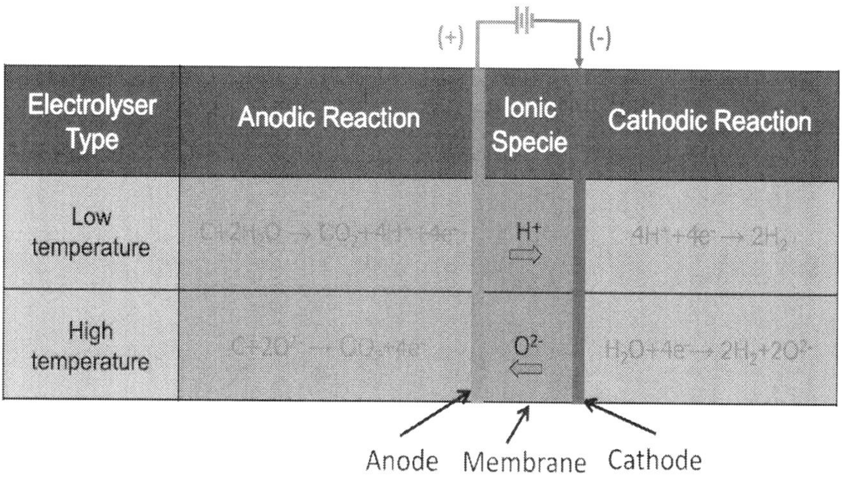

Figure 5: Electrochemical reactions involved in low and high temperature carbon-assisted electrolysis process for hydrogen generation.

In addition to a substantial reduction in the electric energy input by the involvement of carbon, this concept for hydrogen generation combines all three steps mentioned above for hydrogen from NG or coal in a single reactor. The operating temperature is expected to be low (for proton conducting electrolyte membrane used) with the overall reaction being: $C + 2H_2O \rightarrow CO_2 + 2H_2$. Furthermore, the process would generate pure hydrogen and CO_2 in separate compartments of the electrochemical cell separated by the impervious electrolyte membrane. Thus, the substantial cost and the 20–25% energy penalty for CO_2 capture/separation, as is the case with other routes above, can be avoided. Carbon source can be coal or biomass. All these advantages directly translate into a highly efficient process with low overall cost and substantially reduced CO_2 emissions.

While the hydrogen generation by carbon-assisted electrolysis clearly offers significant advantages, the area is largely unexplored. Most of the investigations have been performed with sulfuric acid as the electrolyte and at temperatures below 100°C (Seehra and Bollineni, 2009; Hesenov et al., 2011;Ewan and Adeniyi, 2013). The current densities achieved are very low due to the slow carbon oxidation kinetics at LTs, and formation of films on the surface (such as illite, siderite, carbonate, etc.) of the coal particles that block the active sites on coal, thus making the reaction unsustainable (Jin and Botte, 2010). The slow kinetics of carbon participation in the electrolysis reaction requires new catalytic electrodes and electrolyte materials for optimum performance. The effect of carbon structure, purity, morphology, catalytic additives on the cell performance also requires a more detailed investigation.

A possible strategy to increase the reaction kinetics and improve the hydrogen production rates is to substantially increase the operating temperature of the carbon-assisted electrolyzer with the use of ceramic electrolytes such as doped zirconia (Figure 5). This has the added advantage that it can further reduce the electrical power requirement as discussed in the HT electrolysis section of this article. The voltage required for HT carbon-assisted electrolysis is significantly lower than that required for the PEM-based system described above with some reports showing that hydrogen can be produced even with no applied voltage (Lee et al., 2011). Although this approach could theoretically have significant advantages in terms of cost per unit hydrogen produced, research in this area is still at a very early stage with little understanding of the mechanisms involved or the stability of materials under these operating conditions (Alexander et al., 2011; Ewan and Adeniyi, 2013). If this technology is to be taken forward, a significant effort would be required to understand the fundamental science before designing a prototype device.

ENERGY CONVERSION TECHNOLOGIES

Fuel Cells—the Next Generation

A wide variety of fuel cell systems of various scales (few W to MW range) are now commercially available and their operating regimes and widely varying performance characteristics have been discussed in the literature (Devanathan, 2008; Giddey et al., 2012; Kulkarni and Giddey, 2013;Badwal et al., 2014). These devices have traditionally been categorized firstly by the type of electrolyte and then by the type of fuel used. Fuel cells can be further categorized by the operating temperature, with polymer electrolyte membrane fuel cells (PEMFC) typically have the lowest operating temperatures below 100°C and SOFCs the highest operating around 800°C or above (Figure 6).

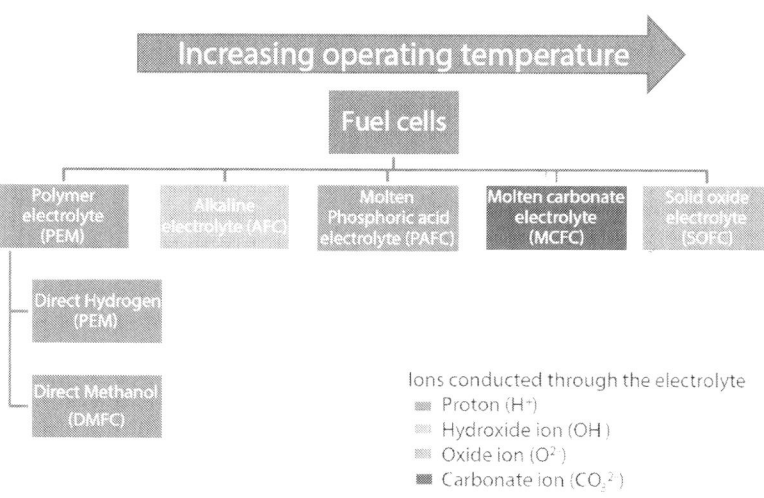

Figure 6: Classification of current commercial or near commercial fuel cell systems.

Conventional Fuel Cells.

The operating temperature in conventional fuel cells is a critical parameter when looking at the system as it defines the type of fuel used, materials choice, end-user application, and electrical efficiency. HT systems (such as molten carbonate and SOFCs) operate at temperatures high enough to allow internal reforming of hydrocarbon fuels. This typically allows these systems to operate with total electrical efficiencies of between 45 and 60%. In contrast the

LT fuel cell systems operating on hydrocarbon fuels must externally reform and clean (removing carbon monoxide) any hydrocarbon fuel used within the system. The operating temperatures of this class of fuel cells is too low to be utilized for reforming hydrocarbon fuels thus leading such systems to have lower electrical efficiencies (around 35–40% total system electrical efficiency when operated on hydrocarbon fuels) when compared to HT systems. Also the PEMFC has a very low tolerance to CO. Intermediate temperature fuel cells (typically operating between 150°C and 350°C) are in general more resilient to fuel impurities and require lower catalyst loadings. This leads to longer operating lifetimes but their electric efficiency is similar to that of LT fuel cells. If low or intermediate temperature fuel cell systems are operated directly on hydrogen, electric efficiencies greater than 50% (with system efficiency over 80% with heat recovery) can be achieved as the fuel processing losses are avoided. Table 1 compares the electrical and system efficiencies of different fuel cell systems operated on reformed hydrocarbon fuels with the values for fuel cells which directly electrochemical oxidize a fuel (Giddey et al., 2012). Any energy from the fuel that is not converted into electrical power is lost as waste heat. A detailed description of how to calculate the total efficiency of a fuel cell system can be found in the following reference (Giddey et al., 2012). In systems where the theoretical efficiency is greater than 100% the fuel cell would require heat input for continuous operation.

Table 1: Theoretical electrical efficiency of fuel cells operated on various fuels with commonly reported system values

Fuel cell type	Fuel	Overall reaction	Operating temperature ($\circ$C)	Theoretical efficiency (%)	Actual system efficiency (%)	
				Electric	Electric	CHP
PEMFC	H_2	$H_{2(g)} + 1/2 O_{2(g)} = H_2O_{(l)}$	60–80	83	45–50	80–90
PEMFC	NG	$CH_{4(g)} + 2O_{2(g)} = CO_{2(g)} + 2H_2O_{(l)}$	60–80		35–40	80–90
DMFC	CH_3OH	$CH_3OH_{(l)} + 1_{1/2}O_{2(g)} = CO_{2(g)} + 2H_2O_{(l)}$	20–60	97	20–25	n/a
AFC	H2	$H_{2(g)} + 1/2 O_{2(g)} = H_2O_{(l)}$	70	83	45–60	n/a
PAFC	NG	$CH_{4(g)} + 2O_{2(g)} = CO_{2(g)} + 2H_2O_{(g)}$	200		40	90
SOFC	NG	$CH_{4(g)} + 2O_{2(g)} = CO_{2(g)} + 2H_2O_{(g)}$	600–1000	92	45–60	90
MCFC	NG	$CH_{4(g)} + 2O_{2(g)} = CO_{2(g)} + 2H_2O_{(g)}$	650	92	45–55	90
DCFC	Carbon	$C_{(s)} + O_{2(g)} = CO_{2(g)}$	500–1000	100	70–80	90

The maximum electric efficiency of a fuel cell system operating on a reformed fuel, in general, is significantly lower than the theoretical maximum where fuel is directly oxidized in the electrochemical reaction of the fuel. This is because all current fuel cells operate on either pure H_2 or (at HT) a mixture of CO and H_2. These fuels are produced, in general, via the reforming or gasification of a hydrocarbon fuel. Reforming of any readily available hydrocarbon fuel requires significant energy input. This is particularly detrimental when an external reformer and fuel processer is used (as is mostly the case for low and intermediate temperature fuel cell systems) because none of the low grade waste heat produced via the fuel cell reactions can be used for reforming. External reforming and fuel processing is a requirement for all LT systems as these systems operate significantly below the temperature required for external reforming (around 500°C). Higher temperature systems can use waste heat from the reactions within the fuel cell to reform the incoming fuel. This results in significantly higher electrical efficiencies being reported for HT commercial systems that operate in this manner (45–60%).

There are two strategies being pursued in order to further increase the efficiency of HT fuel cells operated on gaseous hydrocarbon fuels. The first is to improve the thermal coupling between the fuel cell and the reforming reactions. This is achieved in practice by reducing the physical distance between the zone where the reforming reactions occur and the fuel cells themselves with the ideal being the direct injection of the fuel into the anode chamber. This strategy has a number of technical challenges associated with the instability of hydrocarbon fuels at HTs. These fuels typically decompose to carbon (coking) on the anode surface during the HT operation. This carbon formation can be rapid and results in the fuel cell anode being irreparably damaged. It is also common for coking to occur within the pipe work leading into a fuel cell stack blocking the pipes and stopping the fuel supply to the fuel cell. Coking can be avoided if significant amounts of steam or CO_2 can be introduced to the fuel stream, however, this will significantly reduce the efficiency of the system.

An alternative strategy is to use materials that are more resistance to coking (typically ceramic- or Cu-based anodes). If the residence time of the fuel exposed to HT can be reduced and if anode materials which do not catalyze coking reactions can be used, then it is possible to electrochemically oxidize hydrocarbon fuels directly within a fuel cell via a multi-stage process on the surface of the anode. A number of authors have reported direct oxidation of simple hydrocarbon fuels (such as CH_4), however, the practical difficulties associated with supplying an unstable fuel directly to the reaction sites within a fuel cell have meant that this approach has never been successfully demonstrated at any significant scale (Carrette et al., 2005).

The system cost generally increases with increasing operating temperature as more expensive materials must be used within the system to withstand the harsher operating environment. Detailed reviews of the status of current high, intermediate and low temperature fuel cells are available in the references (Carrette et al., 2005; Devanathan, 2008; Giddey et al., 2012; Kulkarni and Giddey, 2013;Badwal et al., 2014).

Although fuel cell systems are becoming increasingly commercially available there are still sufficient technical challenges that need to be overcome before the mass adoption of fuel cell technology can take place. These challenges relate to lifetime, cost, and suitable fuel supply (for low or intermediate temperature systems). Significant progress is being made through careful engineering of systems to alleviate a number of the issues, including the development of new materials with longer lifetimes, development of materials to allow transport and storage of hydrogen, low cost fabrication technologies for cell and system components and miniaturized fuel processing units for use with LT fuel cells. These advancements are incrementally increasing the appeal of fuel cell systems, however, new developments are required to make the revolutionary advancements necessary to allow fuel cells to begin to displace a significant fraction of conventional power generation capacity.

There is no one fuel cell technology that stands out as being a clear leader in terms of technology maturity or technical superiority. In general the main focus is to develop more fuel flexible systems that can operate on a wider range of fuels at increased electrical efficiency. The requirement for increased efficiency is driving research and development away from systems requiring fuel pre-processing toward systems where the fuel is directly electrochemically oxidized or where the fuel is directly fed to the anode chamber within a fuel cell. This is because this allows the maximum transfer of chemical energy to electrical energy with any waste (thermal) energy from the operation being available to either maintain the operating temperature of the device or used directly in the chemical or electrochemical reactions within the fuel cell chamber. In addition, there is also an increased interest in lowering the operating temperature of fuel cells to reduce overall system cost whilst extending the life of the fuel cell.

Emerging Fuel Cell Technologies

Emerging fuel cell technologies do not fit comfortably within traditional fuel cell categories in particular due to the varied nature of the fuel handling systems and the move away from conventional electrolytes. This leads to them being better defined by the state/ type of the fuel rather than electrolyte chemistry as this is more relevant to the system design and the end use application of the system. Examples of this are direct methanol or ethanol or carbon fuel cells. This classification system is not ideal as there is significant ambiguity as to in which class a fuel cell should reside. In particular, depending on the operating temperature or pressure, the fuel may be either a gas or a liquid. Figure 7 shows a broad fuel-based classification of different fuel cells currently being investigated and is color coded to give an indication of the potential end user applications for each fuel cell type.

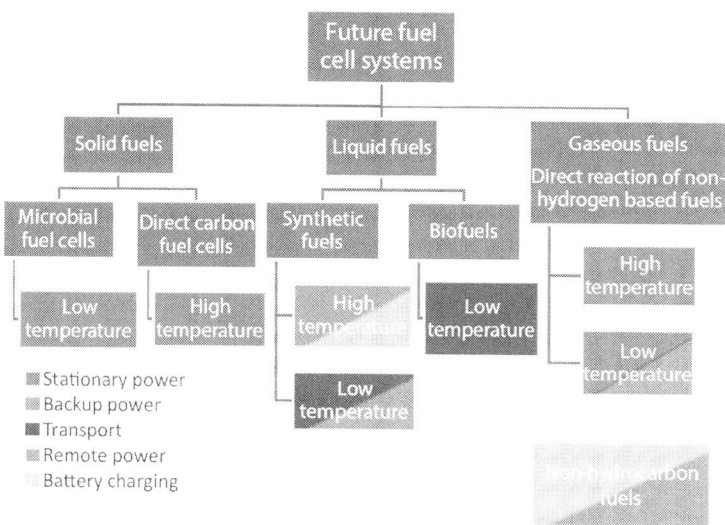

Figure 7: Classification of future fuel cell systems.

Systems based on solid fuels have the attraction that these fuels are often low cost and more abundant than liquid or gaseous fuels. The gaseous fuels have the advantage of being reasonably abundant and can be easily transported over long distances through conventional pipe networks. Liquid fuels are the least abundant of all of the potential fuel sources but are easy to transport and high energy densities make them most suited to transport or mobile applications.

Within the solid fuel class, there are two fuel cell types that could potentially result in a paradigm shift with respect to power generation and application potential: Microbial Fuel Cells (MFC) and Direct Carbon Fuel Cells (DCFC).

Microbial Fuel Cells (MFC)

MFC convert organic material into electrical energy via the microbes' metabolic processes. The use of microbes to produce electric current has been explored since the 1970s but has only recently been revisited for use as a power source for small scale applications as higher power densities are being demonstrated (Rabaey et al., 2009). MFC generally take two forms, membrane reactors and single chamber fuel cells. Within a membrane reactor device, the anode and cathode are separated into two chambers by an electrolyte membrane whereas with single chamber devices both the anode and cathode are in one chamber but separated by organic material. The second class are typically referred to as sediment cells. In both classes of MFC, microorganisms form a biofilm on the surface of the anode and oxidize organic material. These microorganisms then transfer electrons to the anode of the fuel cell either directly (Figure 8A) via micro-pili or indirectly via a mediator (Figure 8B).

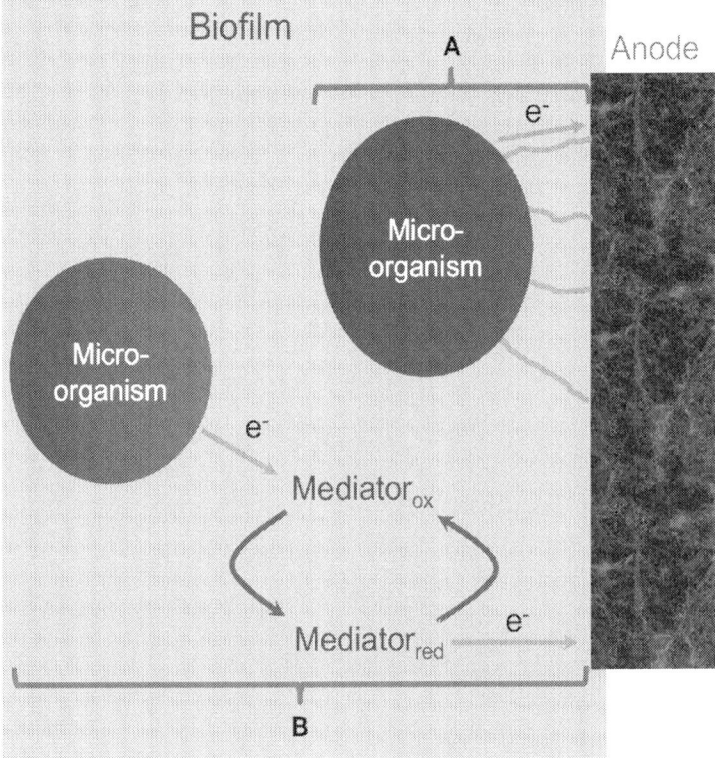

Figure 8: Two modes of operation of a MFC. (A) Direct reaction, and (B) indirect reaction. Figure reproduced from data in Knight et al. (2013).

MFC are considered promising as they operate at or near room temperature and can utilize low grade waste materials such as soils and sediments, waste water and agricultural waste streams that are unsuitable for use in any other power generation technology. The main issue, however, is the very low power density of this type of fuel cells which is typically in the μWcm^{-2} range which is several orders of magnitude below that of other fuel cell types (Rabaey et al., 2009; Knight et al., 2013). Although these fuel cells offer promise in certain low power demand applications, if they are to be adopted at a large scale for such applications, then the power densities need to be increased substantially to at least in the $mWcm^{-2}$ range.

In addition to the absolute performance of MFC's other critical challenges that need to be overcome include faster response times to varying loads, increased voltage stability, increased lifetime, and improved methods of fuel supply to the electrodes. Unlike the majority of other fuel cell types these issues are not fundamentally materials related with the greatest drivers for improvement being novel designs that allow greater mixing of oxidant or fuel with the microbe laden electrodes, improved coupling between the microbes and the electrodes, and selection or modification of the microbes to increase reaction rates at the electrodes. If the activity of the electrodes could be enhanced then further improvements could be obtained via the modifying of the cell design and materials to reduce resistive losses in the electrolyte and electrodes. This could be most easily achieved by reducing the electrolyte thickness or improving the conductivity of the electrodes and/or electrolyte. These improvements are unlike to have a dramatic effect on the performance of MFC's until the activity of the electrodes is increased which would result in higher current passage through the cell. Rabaey et al. (2009) provides a comprehensive technical overview of MFC technology, detailed information regarding the latest understanding of the mechanisms occurring at the electrodes and information on the various designs that are being trialed globally. Carrette et al. (2005) and Knight et al. (2013)provide more generic information on fuel cells with reference Knight et al. (2013)

focussing on some of the design strategies that can be used with respect to utilizing MFC's as practical power generation systems.

Direct Carbon Fuel Cells (DCFC)

Direct carbon fuel cells and fuel cells that directly electrochemically consume hydrocarbon fuels offer many advantages and could potentially compete in many common market sectors to other fuel cell types. The attraction of direct electrochemical oxidation of carbon or gaseous hydrocarbon fuels is that there is the potential to significantly enhance the electrical efficiency of a fuel cell system if the fuel is directly electrochemically reacted rather than gasified or reformed (Table 1).

The DCFC technology has been described in a considerable detail in a recent review article (Giddey et al., 2012). Some of the benefits of the technology, in addition to high efficiency (>65–70% electric, 90% combined heat and electric), include low CO_2 emission, and as the by-product of carbon oxidation is CO_2, its capture costs and energy requirements are very low. Furthermore, if a solid fuel is used (carbon or a high carbon containing hydrocarbon fuel such as coal or biomass chars) then the stability of the fuel becomes less of an issue. These fuels have far higher stability than liquid or gaseous fuel and hence can be fed to the anode surface where they remain stable until oxidized in a chemical or electrochemical reaction.

The DCFC technology is at an early stage of development with a number of different types of DCFC under consideration with a number of groups globally now reporting operation of small stacks (Giddey et al., 2012). Although the electric efficiency is high (>65–70%), reported power densities for these systems, especially once scaled up, are still significantly lower than that of conventional fuel cells operated on gaseous fuels. This is largely due to the reduced surface area for reaction between the anode and the solid fuel that is incapable of infiltrating a porous anode. In order to improve performance a number of groups globally have trialed various strategies to increase the available surface area for reaction. This has included the use of molten metal anodes, molten salt electrolytes, or mixed ionic/electron conducting anode materials (Damian and Irvine, 2012; Giddey et al., 2012;Kulkarni et al., 2012; Jayakumar et al., 2013). A number of these system designs are now in the process of being scaled up with technical issues such as system life, fuel quality, fuel feed, and system cost all still remaining as critical that need to be resolved before these devices can be demonstrated at any significant scale.

As with conventional HT fuel cell systems, the majority of issues currently hindering development of DCFC relate to materials and in particular the way in which materials react with the fuel and other cell components at HTs. In addition to materials issues, there are likely to be an increasing number of challenges relating to fuel handling and processing as this technology matures leading to larger systems being tested for longer periods. Due to the relative immaturity of the field these issues are, as yet, poorly defined.

Dependent on cell design and construction materials issues vary significantly (Giddey et al., 2012). In general reactivity issues are greatest with cell designs that contain molten components in particular molten salts. In cell designs that do not contain solid ion conducting layers, these issues are common with other molten salt fuel cell designs, such as molten carbonate fuel cells, and are relate to the mobility of the electrolyte and its reactivity with other system components (Kulkarni and Giddey, 2013). Although the degradation mechanisms are common with molten carbonate fuel cell designs, the higher operating temperature of DCFC's (typically 800°C vs. 650°C) leads to accelerated degradation rates (Giddey et al., 2012; Kulkarni and Giddey, 2013). The molten salt within the fuel cell can be contained and separated with a dense oxide ion conducting membrane, in this instance the fuel is normally mixed with the molten salt and contained within the anode chamber. This greatly simplifies the issues relating to mobility of the molten components within the fuel cell and results in high power densities but the dense oxide membrane can be rapidly corroded by the molten salt/fuel mixture. Some progress has been made in reducing the reaction rate but this is still seen as a critical issue (Damian and Irvine, 2012; Giddey et al., 2012). A molten metal can operate well as an anode material when a solid fuel is used, however, these metals are likely to be highly reactive toward impurities within the fuels which will accumulate in the anode chamber and result in solidification of the molten metal. These fuel cells are also limited in terms of operating voltage by the reduction potential of the molten metal which can lead to a significant reduction in overall system efficiency (Giddey et al., 2012; Jayakumar et al., 2013). A fuel cell design, where a mixed ionic electronic conducting (MIEC) anode is used to shift carbon oxidation reaction from electrode/electrolyte interface to anode/fuel interface, is likely to have the least reactivity issues due to the fact that all fuel cell components are solid state. This makes them less reactive toward fuel impurities and in general more stable. However, these materials have lower ionic conductivity than molten salts and lower electrical conductivities than molten metals leading to MIEC DCFC's having, in general, lower power densities when compared to other DCFC designs. If stable materials with high mixed ionic and electronic conductivities can be identified, this fuel cell system would rapidly evolve as a leading contender as it can utilize many of the materials and design features of the more technologically mature SOFC technology (Giddey et al., 2012; Kulkarni et al., 2012; Badwal et al., 2014).

Small and Portable Fuel Cells

In addition to next generation fuel cell systems that operate on gaseous and solid fuels at ultrahigh efficiencies, there is also a drive to develop small scale or portable power sources. In these systems, device volume and weight, fuel energy density, and ease of transport of the fuel are critical with the overall system efficiency being important but less critical than for stationary power generation. Portable fuel cell systems are generally based on LT PEM fuel cell stacks that operate near room temperature on pure hydrogen with a limited number of systems being developed that are based on either SOFC technology or that are based on PEM systems but that operate directly on methanol/water mixes (Giddey et al., 2012; Badwal et al., 2014). If the fuel cell is to be operated on pure hydrogen then this is normally stored either within a metal hydride or light weight compressed hydrogen cylinder. Other fuels under consideration include bio-fuels such as ethanol, synthetic hydrocarbon fuels such as methanol and non-hydrocarbon fuels such as ammonia (Brown, 2001;Choudhary et al., 2001; Giddey et al., 2013). For fuel cells operated on the non-hydrogen fuels, with the exception of direct methanol fuel cells, a fuel processor is required to convert the fuel into either pure H_2 or a mixture of H_2 and CO with the later only being suitable for use in HT fuel cell systems.

The use of a fuel processor can often greatly increase the complexity of the device but simplifies the storage of the fuel, particularly in the case of liquid fuels which can often have exceptionally high energy densities and low cost in comparison to either batteries or gaseous hydrogen storage solutions. However, due to the stringent requirements relating to the purity of hydrogen, the cost of the fuel processor can often significantly increase the overall cost of the device with the fuel processer potentially being greater than the cost of the fuel cells stack itself. This typically limits fuel cell/fuel processor combinations to applications where the cost per kWh is more critical than the cost per kW as this allows the high cost of the fuel processor to be offset by the much reduced cost of the fuel storage solution. Similarly, any additional weight from the processor can be offset by the far higher energy density of the fuel storage solution.

These small and portable fuel cell systems are being developed for a range of end-user applications including stationary backup generators, battery charging, remote area power, auxiliary power units, soldier packs, portable electronic appliances, and small transport applications. There are an increasing number of these devices now commercially available, however, lack of fuel infrastructure and high cost when compared to battery or battery generator combinations remain key challenges that need to be overcome for this market to expand further. Future fuel cell designs should be able to operate directly on a greater variety of commonly available fuels without the requirement for significant amounts of fuel pre-processing. This should lead to far greater efficiencies and hence lower operating costs of fuel cell power systems when compared to conventional power generating technologies which are likely to remain lower cost in terms of capital investment in the medium to long term.

ALKALI METAL THERMO-ELECTROCHEMICAL ENERGY CONVERTERS (AMTEC)

The Alkali-Metal Thermo-electrochemical Converter (AMTEC) is an electrochemical device which utilizes heat from a solar or a nuclear source or from combustion of fossil fuels to generate electricity and is an excellent technology for conversion of heat to electricity (Weber, 1974; Cole, 1983; Ryan, 1999; Lodhi and Daloglu, 2001; El-Genk and Tournier, 2004; Wu et al., 2009). The AMTEC is thermodynamically somewhat similar to the Rankine cycle with conversion efficiencies in the 20–40% range, similar to the Carnot cycle. AMTEC devices offer high efficiency for the operating temperature regime and part-load operation independent of size and high power densities around 1 W/cm^2. Some applications of AMTEC devices include dispersed small scale power generation, remote power supplies, aerospace power systems, and vehicle propulsion.

Typically an AMTEC device consists of a sodium or potassium beta alumina as the electrolyte for the transport of Na$^+$ or K$^+$ ions and sodium or potassium metal as the fluid that drives the device. These materials are known to have high ionic conductivity with ionic transport number for Na$^+$ or K$^+$ close to unity (Badwal, 1994). The electrolyte separates the high pressure (>20 kPa) and HT (700–950°C) section of the device from the low pressure (~100 Pa), LT (100–350°C) side of the cell (Weber, 1974;Cole, 1983; Ryan, 1999; Lodhi and Daloglu, 2001; El-Genk and Tournier, 2004; Wu et al., 2009). A schematic of the AMTEC is described in Figure 9 for a system based on sodium as the working fluid. The liquid metal is supplied to one side of the solid electrolyte. With heat provided from an external source, the liquid metal is evaporated and typically Na vapors are present at the porous anode/dense electrolyte interface at a high pressure (>20 kPa). At the cathode, the Na vapor pressure is reasonably low (~100 Pa).

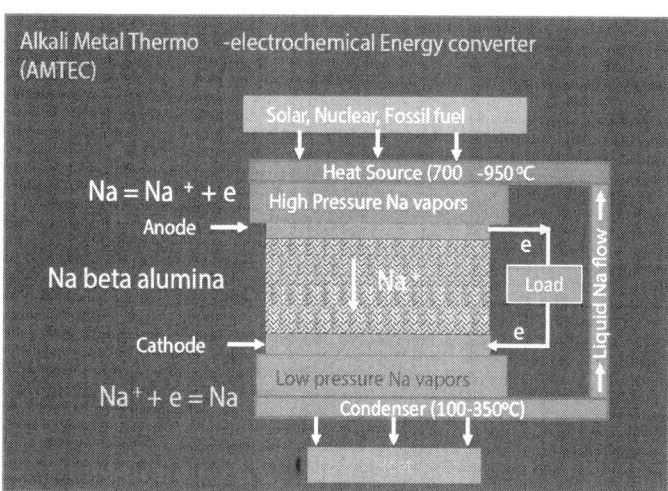

Figure 9: The operating principle of an Alkali Metal Thermo-electrochemical Energy converter (AMTEC).

Due to differential partial pressure of Na across the Na⁺ conducting electrolyte membrane, a voltage signal develops (typically in the V range). When the cell is connected through an external load, under this potential gradient, the sodium in the vapor form is ionized to form Na⁺ ions and electrons at the anode/electrolyte interface. The Na⁺ ions are transported across the electrolyte membrane and recombine with electrons at the cathode (low pressure side) thus producing electricity. The sodium vapors are condensed and cycled back to the anode side for revaporization and the cycle is repeated. A number of cells are connected in series/parallel arrangement to construct a module to meet power requirements of an application. There are no moving parts within the cell and therefore the device has low maintenance requirements. The AMTECs are modular in construction and in many respects have common features with batteries and fuel cells.

The technology has been under development since late 1960s with initial effort going into liquid sodium anode based devices. However, due to low cell voltage and power density, more recent effort has been directed toward vapor phase anode or vapor fed liquid anode systems with significant advances made in the development and manufacturing with performance of multi tube modules demonstrated for several thousand hours of operation (Wu et al., 2009). AMTEC systems in the 10s of kW range have been developed and deployed for space applications (Weber, 1974; Cole, 1983; El-Genk and Tournier, 2004; Wu et al., 2009).

Despite the simple operating principle of the AMTEC device and demonstration of the technology at multi kW level, the technology is quite complex with several severe issues still contributing to the cost, system efficiency, and lifetime. These include: stability of electrodes, electrolyte, and other materials of construction during operation leading to cell power degradation with time; sodium fluid flow management including heat removal during condensation on the cathode side to heat input on the anode side; power controls; system design; and low cost technology up-scaling. The electrode materials play a critical role for charge exchange at the electrode–electrolyte interface and contribute significantly to cell performance (efficiency and degradation). A number of different materials ranging from metals to ceramics or composites of metals and ceramics have been tried with varying degrees of success (Wu et al., 2009). The electrolyte material is also prone to changes in electrical, chemical, and thermo-mechanical properties with extended operation leading to degradation with time. Thus, although the technology offers many advantages for an extensive range of applications, further improvements to lifetime, reliability, power density, and efficiency are required.

ENERGY STORAGE

The implementation of energy storage for applications including transportation and grid storage has strong commercial prospects. A number of market and technical studies anticipate a growth in global energy storage (Yang et al., 2011; Akhil et al., 2013). The main forecasted growth of energy storage technologies is primarily due to the reduction in the cost of renewable energy generation and issues with grid stability, load leveling, and the high cost of supplying peak load. Additionally, the demand for energy storage technologies such as rechargeable batteries for transportation has also added to the forecasted growth. A number of battery technologies have been commercialized and additionally a large number are still under development.

RECHARGEABLE METAL-AIR BATTERIES

The development of nearly all electrically powered devices has closely followed that of the batteries that power them. By way of example, the size and form of today's mobile phones is largely determined by the dimensions of the lithium-ion cells that have the required capacity. Electric vehicles for passenger transportation are an obvious exception. Here, the batteries and electric drive are replacing systems based on liquid-fuel fed combustion engines that provide levels of performance (acceleration, distance between refueling, etc.) which are taken for granted by the motoring public. There is general reluctance by vehicle owners to embrace electric cars offering considerably less all-round performance. This is the main factor that drives researchers to look well-beyond current lithium-ion technology to a range of new metal-air batteries. By virtue of removing much of the mass of the positive electrode, metal-air batteries offer the best prospects for achieving specific energy that is comparable with petroleum fuels.

Lithium-Air (Oxygen)

In its simplest form, the lithium-air cell brings together a reversible lithium metal electrode and an oxygen electrode at which a stable oxide species is formed. There are two variants of rechargeable

Li-air technology—a non-aqueous and an aqueous form, both of which offer at least ten times the energy-storing capability of the present lithium-ion batteries (Girishkumar et al., 2010; Bruce et al., 2011; Kraytsberg and Ein-Eli, 2011; Imanishi and Yamamoto, 2014). Figure 10 provides a schematic view of the two versions. In both, the cathode is a porous conductive carbon which acts as the substrate for the reduction of oxygen, while the anode is metallic lithium. For the non-aqueous system, the reduction of oxygen ends with formation of peroxide, so that the overall reaction follows Equation (1).

$$2\mathrm{Li(s)} + \mathrm{O_2(g)} = \mathrm{Li_2O_2(s)} \tag{1}$$

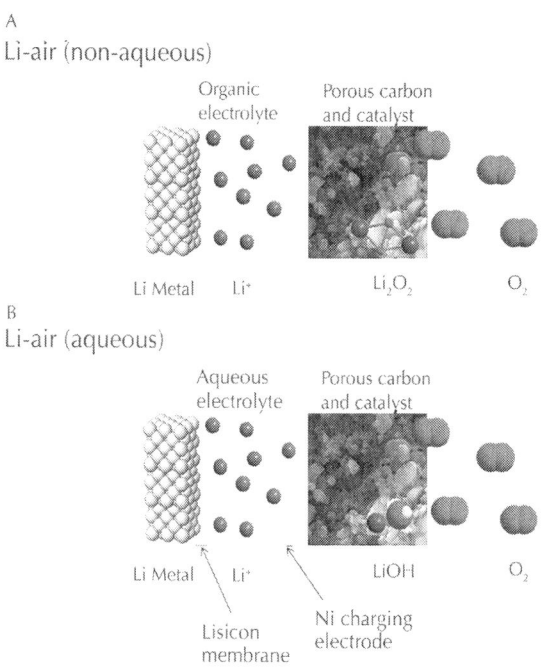

Figure 10: Schematic representation of two contemporary versions of the lithium-air battery—(A): non-aqueous version, similar to Li-ion; (B): aqueous, with Li⁺-permeable membrane protecting the lithium anode.

A cell based on this reaction has an open circuit voltage of 2.96 V and operates at specific energy values ranging between 3460 and 11,680 Wh kg^{-1}. During discharging, the cell draws in oxygen and thereby gains mass, while it loses mass during charging, so that specific energy reaches a maximum when fully charged.

In the aqueous form of lithium-air battery, water is involved in the reduction of oxygen, while the lithium electrode must be protected from reaction with water, usually by means of a lithium-ion-conducting solid electrolyte such as LISICON. Typically the electrolyte solution is a saturated solution of LiCl and LiOH and the favored reduction product is a hydrated lithium hydroxide, according to Equation (2).

$$4Li(s) + 6H_2O(l) + O_2(g) = 4(LiOH.H_2O)(s)$$

(2)

The involvement of water in the reaction complicates the operation of the cell and degrades the specific energy which is theoretically around 2000 Wh kg^{-1} and varies over ~100 Wh kg^{-1} with state-of-charge (Imanishi and Yamamoto, 2014). While this is still an impressive level of performance, the main problem with the aqueous form of lithium-air is the difficulty of maintaining separation of lithium metal from the aqueous medium. Most of the Li$^+$-conducting solids tried to date do not have sufficient long-term stability against aqueous solutions. In addition they contribute significantly to cell impedance—reducing the thickness of this protective layer ameliorates this effect but is limited by the poor mechanical strength of very thin layers. For these reasons, most research effort in lithium-air batteries is focusing on the non-aqueous form.

Clearly a key aspect to the realization of the very high specific energy of lithium-air battery is that the lithium metal anode can be made to operate safely and at full utilization. Many early studies used the organic carbonate electrolytes from lithium-ion battery technology, until it was eventually discovered that these compounds (ethylene carbonate, propylene carbonate, etc.) were being oxidized during the charging phase, with the liberation of carbon monoxide and carbon dioxide. Solvents with ether functionality have since taken precedence given that they are more stable during charging and also less likely to promote the growth of dendritic morphologies at the lithium electrode (Abraham and Jiang, 1996). Nevertheless, both carbonates and ethers are flammable which ultimately makes these devices hazardous under conditions where they become hot. It is not surprising therefore that interest has turned to the use of ionic liquids, which are essentially non-volatile and able to dissolve appreciable concentrations of most lithium salts. In addition, lithium electrodes operate with a high degree of reversibility in a range of low viscosity ionic liquid media, without the formation of dendrites, due to the formation of a durable solid electrolyte interphase (SEI) on lithium (Howlett et al., 2004). An increasingly attractive option to the metallic lithium electrode is to use one of the high capacity lithium host materials, notably silicon which offers the prospect of almost 2000 mAh g^{-1} by accessing the full available storage limit (based on $Li_{4.4}Si$).

The positive electrode of a lithium-air cell represents a complex challenge in that it must provide for: (i) access to oxygen; (ii) wetting by the electrolyte; and (iii) displacement by reaction products. While allowing access to oxygen, the electrode must be able to block access to water, carbon dioxide, and nitrogen, which will all react with the electrode materials and/or products of reaction at the electrodes. The properties of the main product of discharge, lithium peroxide, Li_2O_2, also pose a number of problems with regard to cell longevity. First, it is an insulating solid, which means that conditions must be adjusted to prevent the formation of massive deposits during discharging. Second, lithium peroxide is a strong oxidant that tends to react with electrolyte components, including any adventitious water, to form irreversibly a variety of materials that severely degrade the lifetime of a Li-air cell.

In the last few years, researchers have been able to extract something close to the high levels of performance that the lithium-air system offers, but only for brief periods before rapid capacity loss occurs. The reversibility of oxygen reduction is still the key issue (Mo et al., 2011), and even when conditions are adjusted to promote chemical reversibility, there is a large overvoltage associated with charging which will ultimately work against developing fast-charging procedures. Accordingly, there is still considerable investigation required into the exact mechanism of oxygen reduction, and the oxidation of a range of oxide species, with the aim of greatly improving the energetics of these processes.

Sodium-Air (Oxygen)

The reversible sodium electrode is well-known in the history of battery development as it is featured in some of the very earliest examples of high performance secondary batteries. Both the sodium-sulfur and the Zebra (sodium-nickel chloride) systems employ molten sodium electrodes which give reversible behavior at values of potential that are sufficiently negative for useful device voltages (Ellis and Nazar, 2012). Recently, the sodium electrode has again become the focus of attention, now coupled with an oxygen electrode in the sodium-air cell. This, like all metal-air systems, benefits in energy terms from the inherently lightweight air-breathing cathode and offers theoretical values of specific energy that range from 1105 to 2643 Wh kg^{-1}, depending on the state-of-charge. These numbers are derived from the overall cell reaction shown in Equation (3).

$$Na(s) + O_2(g) = NaO_2(s)$$

(3)

The identification of the superoxide as the main product of reduction has been verified experimentally (Hartmann et al., 2013) and although a basic thermodynamic treatment indicates that it is not the favored product, Ceder's Group has shown that when the discharge products are nanostructured, the surface energetics make the superoxide the preferred product phase (Kang et al., 2014).

Many of the limitations on performance of the air cathode in Li-O_2 cells also define the behavior of this electrode in Na-O_2 cells. The use of carbonate and ether electrolyte solutions has been hampered by problems of insufficient stability during charging (Hartmann et al., 2013). While the preferential formation of sodium superoxide during discharging clearly lowers the overpotential associated with charging, it is not clear whether this compound will be stable on the longer timescale of a typical device service life, or whether the discharge product will gradually be converted to the more stable, and less easily recharged, sodium peroxide.

While the molten sodium electrode offers many advantages in terms of electrochemical characteristics, reality for rechargeable energy storage devices demands that maximum performance is delivered at ambient temperature. What is known of the behavior of solid sodium electrodes in conventional battery electrolytes suggests that it readily generates dendritic morphologies thereby posing a significant risk to further development of this battery technology. By analogy with lithium electrochemistry, it seems likely that more attention will be given to examining the behavior of sodium in ionic liquid electrolytes, in an attempt to replicate the benefits of generating a protective SEI in a medium that is inherently safer with respect to volatility and reactivity.

Although it is very early in the development cycle for sodium-air batteries, there are sound reasons for pursuing further progress. The relative abundance of sodium, compared with lithium, is perhaps the most obvious, and the fact that sodium is close to lithium in terms of mass and electrochemical potential also strengthen the case. Continued larger efforts to develop positive electrode substrates for other metal-air systems (notably lithium) will directly benefit the sodium-air positive electrode. With research already appearing on non-volatile sodium ion-conducting electrolytes based on ionic liquids, it would seem that the main issues holding back the development of sodium-air batteries are now being addressed.

Lithium-Sulfur Batteries

A positive electrode comprised solely of elemental sulfur has a theoretical specific capacity of 1672 mAh g^{-1}. Assuming an equivalent amount of lithium for the negative electrode, complete reaction of Li and S to form Li$_2$S, and an average discharge potential of 2.2 V per cell, the electrode specific energy for Li-S is 2600 Wh kg^{-1} (Bruce et al., 2012; Manthiram and Su, 2013; Song et al., 2013). The overall discharge reaction, in its simplest form, is given in Equation (4), and a schematic view of the components and their role is provided in Figure 11.

$$2\text{Li}(s) + \text{S}(s) \rightarrow \text{Li}_2\text{S}(s)$$

(4)

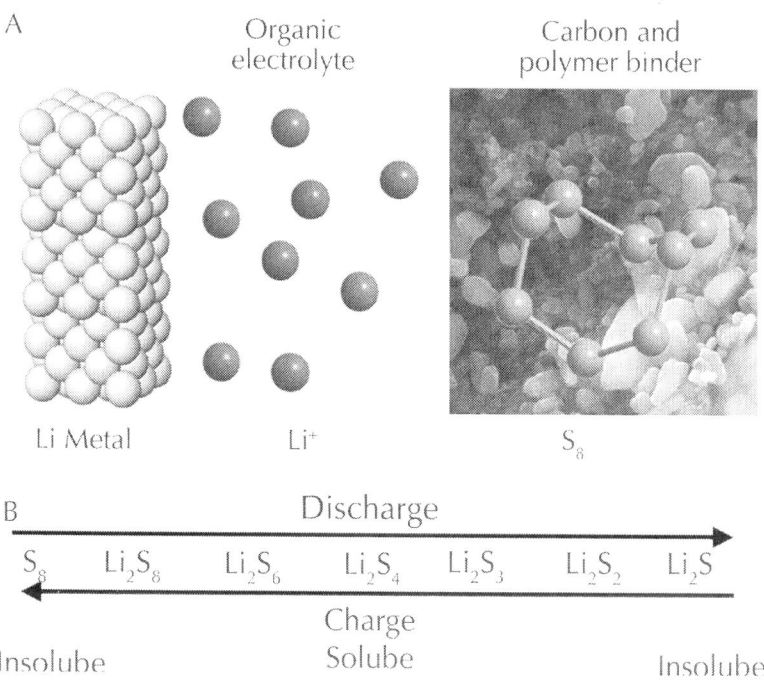

Figure 11: (A) A schematic view of the lithium-sulfur cell. (B) Summary of reactions that define Li-S and their relationship with solubility.

Fully packaged, it is expected that Li-S batteries in real life will operate at up to 700 Wh kg^{-1}. This level of performance places lithium-sulfur well-clear of existing battery systems, and many view it as a logical intermediate step to the lithium-air battery. In many ways, lithium-sulfur also poses a set of mid-level challenges to battery researchers.

While not sharing the full range of difficulties of the air electrode, the sulfur electrode still represents a complex electrochemical system in which elemental sulfur, in the form of S_8 molecules, is successively reduced through a sequence of polysulfide dianions (Bruce et al., 2012). The solubility of the lithium salt of each successive reduction product decreases appreciably, with the end discharge product, Li_2S, being virtually insoluble in common organic electrolyte media. Overlaying this is the generally labile nature of exchange between intermediate members of the polysulfide series, which has the undesirable consequence of allowing significant loss of efficiency through a redox shuttle phenomenon (Manthiram and Su, 2013). As a result of these solution-based issues, most research groups strive to minimize the solubility of polysulfides in the electrolyte.

As it happens, however, controlling the solubility of sulfur and its reduction products is not sufficient on its own to stabilize the performance of the lithium-sulfur battery. It is now clear that the positive electrode, which is the mechanical support for sulfur, must not only be conductive, but also mesoporous, to maximize electrode area within dimensions that do not restrict ion diffusion, and to incorporate surface functionality that acts to adsorb polysulfides so as to enhance the retention of discharge products within the positive electrode. With this knowledge, the design of sulfur positive electrodes now typically incorporates additives such as mesoporous silica, to enhance retention of polysulfides within the electrode, and nano-structured polymer films with chemical functionality to restrict the flow of sulfur species out of the electrode.

In the presence of sulfur and polysulfides, the use of lithium metal as the negative electrode is more complicated than in other lithium battery systems due to a range of interactions between metallic lithium, sulfur species, and electrode-stabilizing additives such as lithium nitrate (Aurbach et al., 2009). Helping to provide greater control over the behavior of the lithium electrode is the increasing trend to incorporate ionic liquids in Li-S electrolyte blends. Here it is the fluorosulfonyl imide anions (either FSI or TFSI), which contribute to the formation of a stable SEI, that provide the basis for safe, dendrite-free operation of the lithium negative electrode. More recently, it has also been discovered that lithium ion transport characteristics can be greatly enhanced, while at the same time suppressing the solubility of polysulfide species, by increasing the concentration of the lithium salt to unprecedented levels (>5 M).

Despite the high degree of chemical complexity inherent to the lithium-sulfur battery, there are strong signs that the issues which have thwarted progress are now being brought under control, mainly through the tailoring of electrode and electrolyte materials to deal with specific aspects of performance. At the same time, it is interesting to note that the development of lithium-sulfur battery technology also seems likely to give rise to a successful all-solid component version, due to the advent of a family of high-lithium-ion-conducting ceramic sulfides (Kamaya et al., 2011).

FLOW BATTERIES

A flow battery is a rechargeable battery where the energy is stored in one or more electroactive species dissolved into liquid electrolytes. The electrolytes are stored externally in tanks and pumped through electrochemical cells which convert chemical energy directly to electricity and vice versa, on demand. The power density is defined by the size and design of the electrochemical cell whereas the energy density or output depends on the size of tanks. With this characteristic, flow batteries can be fitted to a wide range of stationary applications. Originally developed by NASA in the early 1970's as electrochemical energy storage systems for long-term space flights, flow batteries are now receiving attention for storing energy for durations of hours or days. Flow batteries are classified into Redox flow batteries and hybrid flow batteries.

Flow batteries have the advantages of low cost devices, modularity, easy transportability, high efficiency and can be deployed at a large scale (Ponce de Leon et al., 2006). The modularity and scalability of these devices means they can easily span the kW to MW range. As a result, their main development at present is focussed on standalone remote area power systems or grid energy storage/support in combination with renewable energy generation (Skyllas-Kazacos et al., 2011).

Redox Flow Battery (RFB)

In redox flow batteries (RFB), two liquid electrolytes containing dissolved metal ions as active masses are pumped to the opposite sides of the electrochemical cell. The electrolytes at the negative and positive electrodes are called negative electrolyte (also referred to as the anolyte) and positive electrolyte (also referred to as the catholyte), respectively. During charging and discharging the metal ions stay dissolved in the fluid electrolyte; no phase change of these active masses takes place. Negative and positive electrolytes flow through porous electrodes, separated by a membrane which allows protons to pass through it for the electron transfer process. During the exchange of charge a current flows over the electrodes, which can be used by a battery-powered device. During discharge the electrodes are continually supplied with the dissolved active masses from the tanks; once they are converted, the resulting product is removed to the tank.

Various redox couples have been investigated and tested in RFBs, such as a Fe-Ti system, a Fe-Cr system, and a polyS-Br system. The vanadium redox flow battery (VRFB) has been developed the furthest; it has been piloted since around 2000 by companies such as Prudent Energy (CN) and Cellstrom (AU). The VRFB uses a V^{2+}/V^{3+} redox couple as the negative pair and a V^{5+}/V^{4+} redox couple in mild sulfuric acid solution as the positive pair. The main advantage of this battery is the use of ions of the same metal on both sides.

Although crossing of metal ions over the membrane cannot be prevented completely (as is the case for every Redox flow battery), in VRFBs the only result is a small loss in energy. In other RFBs, which use ions of different metals, the crossover causes an irreversible degradation of the electrolytes and a loss in capacity. The VRFB was pioneered at the University of New South Wales, Australia, in the early 1980s (Skyllas-Kazacos et al., 2011).

Hybrid Flow Battery (HFB)

In a hybrid flow battery (HFB) one of the active masses is internally stored within the electrochemical cell, whereas the other remains in the liquid electrolyte and is stored externally in a tank. Therefore, hybrid flow cells combine features of conventional secondary batteries and redox flow batteries: the capacity of the battery depends on the size of the electrochemical cell. Typical examples of a HFB are the Zn-Ce (Fang et al., 2002; Clarke et al., 2004; Ponce de Leon et al., 2006; Reddy, 2011) and more commonly the Zn-Br$_2$ system (Lim et al., 1977; Lex and Jonshagen, 1999; Ponce de Leon et al., 2006; Reddy, 2011). In both cases the negative electrolyte consists of an acid solution of Zn^{2+} ions. During charging Zn is deposited at the electrode and on discharging Zn^{2+} goes back into solution. In the case of the Zn-Br systems the electrode reactions are shown below:

During discharge, the zinc in the anode is oxidized:

So that the overall reaction is:

$$Zn \overset{\text{discharge}}{\underset{\text{charge}}{\rightleftarrows}} Zn^{2+} + 2e \qquad -0.763\,V$$

(5)

At the cathode bromine is reduced, to bromide, Br⁻,

$$Br_2 + 2e \underset{charge}{\overset{discharge}{\rightleftharpoons}} 2Br^- \qquad +1.087\,V$$

(6)

$$Zn + Br_2 \underset{charge}{\overset{discharge}{\rightleftharpoons}} Zn^{2+} + 2Br^- \qquad +1.850\,V$$

(7)

The two electrode chambers of each cell are separated by a membrane (typically a microporous or ion-exchange variety). This helps to prevent bromine from reaching the positive electrode, where it would react with zinc, causing the battery to self-discharge. To further reduce self-discharge and to reduce the vapor pressure of bromine, complexing agents are added to the positive electrolyte. These react reversibly with the bromine to form an aqueous solution and reduce the free Br_2 in the electrolyte. The working electrodes in the $Zn-Br_2$ battery are based on carbon-plastic composites.

Various companies are working on the commercialization of the $Zn-Br_2$ hybrid flow battery, which was developed by Exxon in the early 1970s. In the United States, ZBB Energy and Premium Power sell trailer-transportable $Zn-Br_2$ systems with unit capacities of up to 1 mW/3 mWh for utility-scale applications. Some 5 kW/20 kWh systems for community energy storage are in development as well. In Australia, Redflow Ltd. has developed a $Zn-Br_2$ system for electrical energy storage applications. $Zn-Br_2$ batteries can be 100% discharged every day without being damaged and this can be repeated for over 2000 cycles.

Flow Battery Future Prospects

In addition to the V- and Br_2-based systems, a number of alternative chemistries are also being investigated. The reason for this is that the new applications for these devices, such as electricity grid integration, require that the performance, in particular the volumetric energy density is increased. There are a number of challenges still to be overcome to achieve this goal. Firstly, electrode development should focus on porous and catalytic electrodes which allow high electrolyte linear flow velocities to enhance rate capability (Ponce de Leon et al., 2006). Secondly, the engineering of the device also requires attention in the areas of reactor design, electrode materials to enhance catalysis (Ponce de Leon et al., 2006), membrane performance (Ponce de Leon et al., 2006) to reduce migration of active species and finally the large scale engineering (Ponce de Leon et al., 2006) to allow for up-scaling of the technology for very large installations with focus on minimizing maintenance and increasing life. Two very good review articles by Ponce de Leon et al. (2006) andSkyllas-Kazacos et al. (2011) have given a good overview of the development and challenges of flow batteries. Table 2 summarizes the range of different flow battery chemistries which have been previously reported.

Table 2: Summary of flow battery chemistries reported in the recent literature

System	Electrode reactions	Electrolyte	OCP	References
All-vanadium	Negative electrode: $V^{3}+ + e^- \rightarrow V^2+$ Positive electrode: $VO^{2+} + H_2O - e^- \rightarrow VO_2^+ + 2H^+$	1.6–2M vanadium sulfate in sulfuric acid in both half-cells	1.6	Skyllas-Kazacos and Grossmith, 1987; Skyllas-Kazacos, 2009; You et al., 2009; Jia et al., 2010; Skyllas-Kazacos et al., 2010
Vanadium bromine	Positiveelectrode: $2VBr^3 + 2e^- \rightarrow 2VBr^2 + 2Br^-$ Negativeelectrode: $2Br^- + Cl^- \rightarrow ClBr_2 + 2e^-$	1–3M vanadium bromide in 7–9M HBr plus 1.5–2M HCl in both half-cells	1.4	Skyllas-Kazacos, 2003; Skyllas-Kazacos et al., 2010
Magnesium-vanadium	Positiveelectrode: $Mn(II) \rightarrow Mn(III) + e^-$ Negativeelectrode: $V(III) + e^- \rightarrow V(II)$	Positive half-cell: 0.3M Mn(II)/Mn(III) in sulfuric acid) Negative half-cell: V(III)/V(II) in 5M sulfuric acid	1.66	Xue et al., 2008
Vanadium cerium	Positiveelectrode: $Ce^{3+} \rightarrow Ce^{4+} + e^-$ Negativeelectrode: $V^{3+} + e^- \rightarrow V^{2+}$	Positivehalf-cell:600mlof0.5M Ce(III) in1MH$_2$SO$_4$ Negativehalfcell:600mlof 0.5MV(III)in1MH$_2$SO$_4$	1.05	Paulenova et al., 2002; Xia et al., 2002; Leung et al., 2011a

Vanadium glyoxal(O_2)	Positive electrode: [OC]$_{RE}$ + $H_2O \rightarrow$ [OC]$_{OX}$ + $2H^+$ + $2e^-$ ([OC]$_{RE}$ = organic reductive materials and [OC]$_{OX}$ = electro-oxidized organic products). Negative Electrode: V^{3+} + $e^- \rightarrow V^{2+}$	Positive half-cell:50ml glyoxal–HCl solution of different concentration Negative half-cell:1–2MV(III) + 3M H_2SO_4 solution	1.2	Wen et al., 2008a
Vanadium cystine(O_2)	Positive electrode: RSSR + Br_2 + $6H_2O \rightarrow 2RSO_3H$ + 10HBr(where RSSR = L-cystine and RSO$_3$H = L-cysteic acid) Negative electrode: V^{3+} + $e^- \rightarrow V^{2+}$	Positive half-cell:0.1M cystine dissolved in HBr aqueous solution of different concentrations Negative half-cell:50ml of 1 MV(III) + 3M H_2SO_4	1.315	Wen et al., 2008b
Vanadium polyhalide	Positive electrode: Br^- + $2Cl^- \rightarrow BrCl_2^-$ + $2e^-$ Negative electrode: VCl_3 + $e^- \rightarrow VCl_2$ + Cl^-	Positive half-cell:1M NaBr in 1.5M HCl Negative half-cell:1M VCl$_3$ in 1.5M HCl	1.3	Skyllas-Kazacos, 2003
Vanadium acetylacetonate	Positive electrode: V(III)(acac)$_3 \rightarrow$ [V(IV)(acac)$_3$]$^+$ + e^-. Negative electrode: V(III)(acac)$_3$ + $e^- \rightarrow$ [V(II)(acac)$_3$]$-$	0.01MV(acac)$_3$/0.5M TEABF4/ CH3CN in both half-cells	2.2	Liu et al., 2009
Vanadium/air system	Positive electrode: $2H_2O \rightarrow 4H^+ + O_2 + 4e^-$ Negative Electrode: $V^{3+} + e^- \rightarrow V^{2+}$	Positive half-cell:H_2O/O_2 Negative half-cell: 2M V^{2+}/V^{3+} solution in 3M H_2SO_4	_1 V for 8h	Hiroko et al., 1994, 1997

Bromine polysulfide	Positive electrode: $3Br^- \rightarrow Br_3^- + 2e^-$ Negative electrode: $S_4^{2-} + 2e^- \rightarrow 2S_2^{2-}$	5MNaBrsaturatedwithBr$_2$ and 1.2MNa$_2$S	1.7–2.1	Remick and Ang, 1984; Zhao et al., 2005; Zhou et al., 2006
Zinc-bromine	Positive electrode: $2Br^- \rightarrow Br_2 + 2e^-$ Negative electrode: $Zn^{2+} + 2e^- \rightarrow Zn^{0(s)}$	1–7.7moldm−3 ZnBr$_2$ with an excessofBr$_2$ with additivessuchas KCl orNaCl	1.6	Remick and Ang, 1984; Zhao et al., 2005; Zhou et al., 2006
Zinc-bromine	Positive electrode: $2Br^- \rightarrow Br_2 + 2e^-$ Negative electrode: $Zn^{2+} + 2e^- \rightarrow Zn^{0(s)}$	1–7.7moldm^{-3} ZnBr$_2$ with an excessofBr$_2$ with additivessuchas KCl orNaCl	1.6	Eustace, 1980; Zhou et al., 2006; Nyugen and Savinell, 2010; Weber et al., 2011
Zinc-cerium	Positive electrode: $2Ce^{3+} \rightarrow 2Ce^{4+} + 2e^-$ Negative electrode: $Zn^{2+} + 2e^- \rightarrow Zn^0_{(s)}$	Anolyte: 0.3MCe2(CO$_3$)$_3$ and 1.3M ZnO in70wt.%methanesulfonic acid catholyte: 0.36MCe2(CO$_3$)$_3$ and 0.9MZnOin995gmethanesulfonic acid	2.45	Eustace, 1980; Zhou et al., 2006; Leung et al., 2011a,b
Soluble lead-acid	Positive electrode: $Pb^{2+} + 2H_2O \rightarrow PbO_2 + 2H^+ + 2e^-$ Negative electrode: $Pb^{2+} + 2e^- \rightarrow Pb(s)$	Soluble lead (II) species in methanesulfonic acid	1.62	Hazza et al., 2005; Zhou et al., 2006; Collins et al., 2010

All-neptu-nium	Positive elec-trode: $Np^{3+} \rightarrow Np^{4+} - e^-$ Negative elec-trode: $NpO_2^{2+} + 2e^- \rightarrow NpO_2^+$	1 M nitric acid solution of 0.05 M neptunium	1.3	Hasegawa et al., 2005; Yamamura et al., 2006a
All-uranium	Positive elec-trode: $U(IV) \rightarrow U(V) + e^-$ Negative elec-trode: $U(IV) + e^- \rightarrow U(III)$	U(VI)/U(V) β-diketonate solution as the catholyte and U(IV)/U(III) β-diketonate solution as the anolyte	1.1	Yamamura et al., 2002, 2004, 2006b; Shirasaki et al., 2006a,b
All-chromi-um	Positive elec-trode: $2[Cr(III)EDTA(H_2O)]^- \rightarrow 2[Cr(II)EDTA(H_2O)]^{2-} + 2e^-$ Negative elec-trode: $2[Cr(III)EDTA(H_2O)]^- + 2e^- \rightarrow 2[Cr(II)EDTA(H_2O)]^{2-}$	0.2M chromium EDTA complex in HCl	2.11	Chieng and Skyllas-Kazacos, 1992; Bae, 2001; Bae et al., 2011
Zinc-air	Positive elec-trode: Propanol oxidation during charging; oxygen reduc-tion during dis-charge. Nega-tive electrode: $Zn(OH)_4^{2-} - 2e^- \rightarrow Zn + 4OH^-$	0.4M ZnO in 6M KOH solution was employed as the catholyte and propanol of different con-centrations in 6M KOH solution was employed as the anolyte	1.705	Wen et al., 2009
Tiron	Positive elec-trode: $[Tiron] - 2^- + 2e^- \rightarrow [Tiron]^-$ Negative elec-trode: $Pb - SO_4^{2-} \rightarrow PbSO_4 + 2e^-$	0.25M Tiron in 3M H$_2$SO$_4$ as ca-thodic active species and the lead electrode as anodic active species	1.10	Xu et al., 2010

Zinc-nickel	Positive electrode: $ZnOOH + H_2O + 2e^- \rightarrow Zn(OH)_2 + 2OH^-$ Negative electrode: $Zn + 4OH^- \rightarrow Zn(OH)_4^{2-} + 2e^-$	Highly concentrated solutions of ZnO in aqueous KOH	1.705	Cheng et al., 2007; Zhang et al., 2008
[Ru(acac)3]	Positive electrode: $Ru(acac)_3 \rightarrow [Ru(acac)_3]^+ + e^-$ Negative electrode: $[Ru(acac)_3] + e^- \rightarrow [Ru(acac)_3]^-$	0.02M ruthenium acetylacetonate with 0.1M tetraethylammonium tetrafluoroborate dissolved in acetonitrile	1.76	Sum and Skyllas-Kazacos, 1985; Chakrabarti et al., 2007
Cr(acac)3	Positive electrode: $Cr(acac)_3 \rightarrow [Cr(acac)_3]^+ + e^-$ Negative electrode: $[Cr(acac)_3] + e^- \rightarrow [Cr(acac)_3]^-$	0.05M Cr(acac)$_3$ and 0.5M TEABF$_4$ dissolved in acetonitrile	3.4	Liu et al., 2010
Iron chromium	Positive electrode: $Fe^{2+} \rightarrow Fe^{3+} + e^-$ Negative electrode: $Cr3^+ + e^- \rightarrow Cr^{2+}$	1M CrCl$_3$ and FeCl$_2$ in 2M HCl in the negative and positive sides of the cell, respectively	1.18	Zhou et al., 2006

An emerging concept for flow batteries is the use of microfluidics to remove the membranes from the system. These devices use laminar interfaces between the positive and negative electrolyte streams to separate the reactants. This approach offers the flexibility that allows the exploitation of a much wider range of chemistries. In the literature, chemistries such as vanadium redox flow batteries

(Salloum and Posner, 2010, 2011) and a hybrid hydrogen-bromine flow battery (Braff et al., 2013) have been reported. Typically, the devices have power capabilities in the 0.25 W/cm2 [borohydride-cerium ammonium nitrate (Da Mota et al., 2012) to 0.795 (hydrogen-bromine flow cell Braff et al., 2013)]. This approach allows high efficiencies in the 90% range to be obtained (Braff et al., 2013). Although, the prospects for membrane-less flow batteries looks promising, significant work is still left to do before these devices can become a commercial reality.

SUPERCAPACITORS

Supercapacitors are electrochemical devices that store energy by virtue of the separation of charge, unlike batteries, which store energy through chemical transformation of electrode materials. Known also as ultracapacitors, supercapacitors continue to develop and mature as an energy storage technology, though somewhat still in the shadow of rechargeable batteries. While the designations "ultra" and "super" reflect the fact that these devices have much higher levels of capacitance than traditional capacitors (so-called "electrolytic capacitors," etc.), a more useful, but less popular, name is "electrochemical double-layer" capacitor, which reflects the origins of the very high values of specific capacitance in the electrochemical double-layer that forms at the electrode-electrolyte interface. On this basis, supercapacitors were originally "symmetrical" devices based on two identical electrodes, each comprised of a network of activated carbon particles (Zhang and Zhao, 2009). The latter material provided the very high levels of surface area that are required to give reasonable values of specific energy. This parameter is still the main problem for supercapacitors as, while their specific power (up to several kW kg^{-1} for complete devices) is unrivaled, most electrical storage applications require more than 10 Wh kg^{-1} of specific energy (usually a great deal more) and supercapacitors generally struggle to store more than 5 Wh kg^{-1} (Burke, 2010).

Figure 12 summarizes the essential characteristics of a supercapacitor in a schematic form. The electrodes in a carbon symmetrical device are identical, although the respective loading of active materials will be adjusted in line with small variation of specific capacitance for the different ions that make up the supporting electrolyte. In early devices, strong aqueous electrolytes (e.g., sulfuric acid, sodium hydroxide solutions) were employed as the very high values of ionic conductivity led to maximum power outputs. The device voltage was however limited to around 1 V and this has a great impact on specific energy, courtesy of the squared relationship between capacitor voltage and energy. In the last decade, developments have focused on non-aqueous electrolytes with which it has been possible to gradually raise device voltages up to around 2.7 V (Burke, 2010). Given that these electrolyte solutions are based on flammable solvents (acetonitrile, propylene carbonate, etc.) some recent efforts have also been devoted to employing low viscosity ionic liquids in making inherently safer supercapacitors.

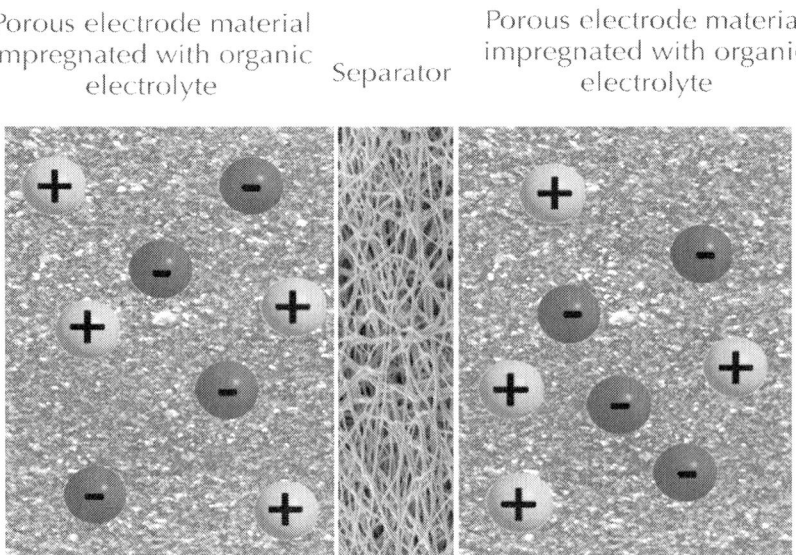

Figure 12: A schematic view of an electrochemical double-layer capacitor, based on a symmetrical carbon-carbon device.

Having noted that "traditional" carbon supercapacitors have not progressed beyond 5 Wh kg^{-1} in specific energy, it is not surprising that research in more recent years has turned to ways of moving to much higher values. The phenomenon known as "pseudocapacitance" has been known for many years and was originally detailed in research on the behavior of ruthenium oxide in aqueous media. Materials like this can be repetitively charged and discharged at 5–10 times the specific energy of carbon supercapacitors via a mechanism that involves movement of highly mobile species (hydrated protons or hydroxide ions, depending on pH) to balance changes of charge of the metal oxide active material. Therefore, these devices have a "Faradaic" basis of operation but are not reversible in a Nernstian sense. With chemical change to the electrode materials clearly involved in the mechanism of charge storage, this inevitably leads to internal stresses during charge–discharge which limits cycle-life to values in the 10,000s—well-short of those obtained with symmetrical carbon devices (up to a million cycles). Nevertheless, impressive gains in specific energy have been made with supercapacitors employing manganese oxides (Wei et al., 2011) and conducting polymers (Snook et al., 2011), both of which draw on pseudocapacitance for energy storage.

As a further progression of the ideas to improve specific energy by introducing materials with greater "energy content" per unit weight, a significant stream of research is now developing so-called "hybrid supercapacitors." These devices incorporate one high surface area carbon electrode and one battery electrode. The latter must be made from a material that is capable of operating at very high rates of charge/discharge, otherwise the performance will not justify the designation as a supercapacitor. To date, the major successes in this field have come with the use of lithium titanium oxide ($Li_4Ti_5O_{12}$, LTO) (Naoi et al., 2013). This material works in this role, where others have failed, because it undergoes virtually no dimensional change between charged and discharged states.

Finally, carbon researchers have been far from idle and there have been marked renewals of interest in carbon supercapacitors due to the development of advanced electrode materials based on nanotubes (Fisher et al., 2013) and graphene (Dong et al., 2013). Both forms of carbon are not only highly conductive and therefore excellent bases for capacitor electrodes, but they also provide excellent supports for chemical modifications with which pseudocapacitance can be incorporated. Graphenes in particular have also been shown to be excellent templates for the mesoporous electrode morhpologies that are essential for balancing the dual requirements of conductivity and ion diffusion. There are strong grounds for confidence in the further development of high power devices with enhanced energy storage capability.

ADVANCED PB ACID BATTERIES

The lead acid battery is one of the most well-known battery technologies to date first demonstrated by Plante in 1859 (Kurzweil, 2010). The lead acid battery is widely used in a variety of applications including automotive, industrial, submarine, and back-up power amongst many others. The lead acid battery is based on the reactions of lead compounds with sulfuric acid in an electrochemical cell. The discharge reaction equations are as shown below.

At the anode:

$$\mathrm{Pb} + \mathrm{SO_4^{2-}} \quad \overset{\text{discharge}}{\underset{\text{charge}}{\rightleftarrows}} \quad \mathrm{PbSO_4} + 2\mathrm{e^-}.$$

$$(8)$$

At the cathode:

$$\mathrm{PbO_2} + \mathrm{SO_4^{2-}} + 4\mathrm{H^+} + 2\mathrm{e^-} \quad \overset{\text{discharge}}{\underset{\text{charge}}{\rightleftarrows}} \quad \mathrm{PbSO_4} + 2\mathrm{H_2O}$$

$$(9)$$

so that the overall reaction equation is:

$$Pb + PbO_2 + 2H_2SO_4 \underset{\text{charge}}{\overset{\text{discharge}}{\rightleftharpoons}} 2PbSO_4 + 2H_2O.$$

(10)

There are two different types of lead-acid batteries. The flooded type is the cheapest and tends to be used in automotive and industrial applications. However, the sealed type, also called valve-regulated lead-acid (VRLA), has been rapidly developed and used in a wide range of applications including hybrid and electric vehicles (Cooper, 2004) and power supplies, such as uninterruptible (UPS) and standalone remote areas power supply (RAPS). The sealed/VRLA type, either with absorptive glass mat (AGM) separators or gelled electrolyte technology, has the advantage of low maintenance (due to acid restriction and oxygen recombination) and easy fit configuration. Both the power and energy capacities of lead-acid batteries are based on the size and geometry of the electrodes. The power capacity can be improved by increasing the surface area for each electrode, which means greater quantities of thinner electrode plates in the battery.

Some advantages of the lead-acid system are its low cost, high power, and most successful recycling rate. One disadvantage of lead acid batteries is usable capacity decrease when high power is discharged. For example, if a battery is discharged in 1 h, only about 50–70% of the rated capacity is available. Other drawbacks are lower energy density and the use of lead, a hazardous material prohibited or restricted in various jurisdictions. Advantages are a favorable cost/performance ratio, easy recyclability and a simple charging technology.

It is due to the power performance drawbacks (Yan et al., 2004) that research into advanced hybrid lead acid systems was instigated. Under high-rate partial state-of-charge cycling applications, the lead acid (VRLA type) battery fails prematurely due to the sulfation of the plates (Catherino et al., 2004; Lam et al., 2004). The negative plates suffer from a progressive build-up of lead sulfate which is difficult to remove during recharge. The accumulation of lead sulfate markedly reduces the effective surface-area so that the plate can no longer deliver and accept the required power.

Two approaches exist to overcome this problem. The first is the connection of a supercapacitor device to take up the power requirements and thereby reduce the sulfation issues faced by the plates. However, this option requires sophisticated electronics and control algorithms which results in a complex device to construct. The second approach, taken by Lam et al., was to combine a supercapacitor and a lead-acid battery within the cell, thereby removing the need for control electronics (Lam and Louey, 2006). In this approach the lead acid cell comprises one lead oxide plate and one sponge lead plate. In addition, the negative lead plate also comprises a carbon-based electrode which uses the lead oxide plate as the counter electrode, thereby forming an asymmetric supercapacitor. Thus during operation, the carbon component of the lead/carbon plate acts to buffer high currents from the lead component thereby allowing an increase in power performance and overall battery life. Overall evaluation of the hybrid battery has demonstrated that the technology has a similar working potential to that of the conventional lead acid battery, low hydrogen gassing rates, higher capacity, long cycle life and can easily be manufactured in existing lead acid battery factories. Further evaluation of this technology with new applications such as grid integration with renewable has demonstrated improved performance and greater cycle life than conventional lead acid batteries.

STORAGE FOR RENEWABLE GENERATION INTEGRATION INTO ELECTRICITY GRIDS

Currently, significant efforts around the world are placed at reducing CO_2 emissions in an effort to mitigate climate change issues caused by excess CO_2 in the atmosphere. In 2011, worldwide 32,600 million tonnes of CO_2 was emitted from the consumption of energy worldwide (International Energy Statistics, 2011). Of the global energy being produced, over 80% is fossil fuels based (WEO, 2012). As a consequence, significant efforts globally have focused on development, demonstration, and deployment of renewable energy generation sources such as wind, solar photovoltaic, tidal, etc. More recently, the efforts have begun to focus on the deployment of energy storage onto electricity grids.

The variable nature of renewable energy generation can create significant issues with grid stability, demand management, etc. When the intermittent generation is less than 15–20% of the overall energy consumption, grid operators are able to compensate for the effects on grid stability (European Commission, 2013). However, an increase in renewable generation above 20–25% creates significant issues especially when demand is also high when combined with intermittency effects from renewable energy generation (U.S. Energy Information Administration, 2014). To minimize these issues and allow greater penetration of renewable generation into the grid, academia, government, grid operators, regulators, and utilities are recommending storage solutions which can stabilize the grid through a combination of energy shifting or direct smoothing. This brings new opportunities for existing storage technologies. However, since the currently available storage technologies, for example batteries, were not initially designed for such purposes these new applications also bring new science challenges to allow the proven and accepted technologies a new lease in life.

Energy storage integration onto the grid encompasses a range of different applications each with their own unique power, energy, and response time requirements. Furthermore, system size, cycle number, and lifetime requirements also vary for the differing applications. A range of different grid applications where energy storage (from the small kW range up to bulk energy storage in the 100's of MW range) can provide solutions and can be integrated into the grid have been discussed in reference (Akhil et al., 2013). These requirements coupled with the response time and other desired system attributes can create limitations on where the energy storage technologies described above can be effectively used.

Figure 13 shows the types of requirements of storage time, power and response time and the types of applications (Chatzivasileiadi et al., 2013). Differing technologies have different power and energy performance characteristics and therefore the application limitations of different technologies are quite obvious in Figure 13. Clearly based on the data some systems will not be suitable for power quality type of applications whilst other would not be suitable for bulk long-term storage type of applications. The performance characteristics of selected energy storage technologies are described in more detail in Table 3 (Chatzivasileiadi et al., 2013).

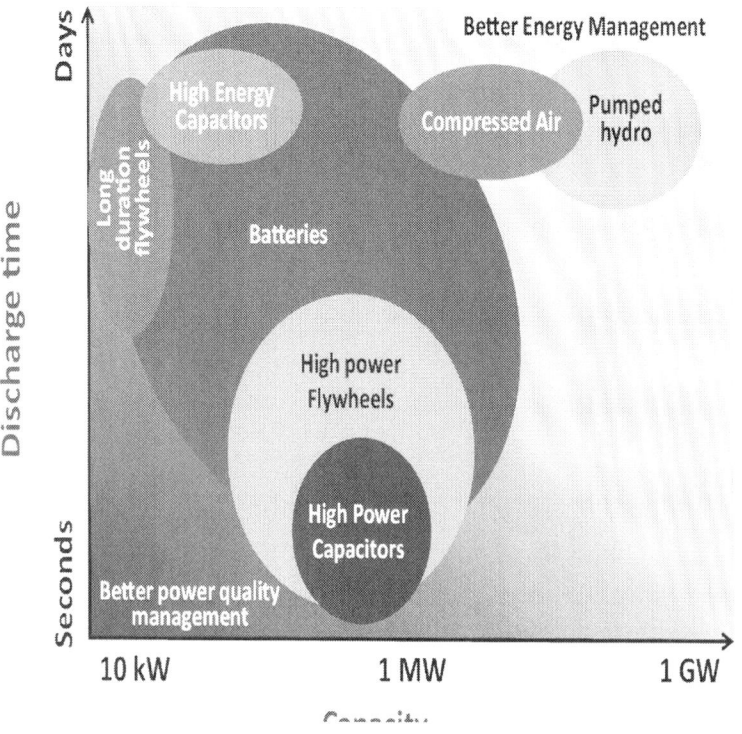

Figure 13: Approximate representation of characteristics of different storage technologies. Some types, especially "batteries," encompass or overlap many technologies within the general shape. Redrawn from the data in Chatzivasileiadi et al. (2013).

Table 3: Characteristics of different battery energy storage technologies are summarized (adapted from Chatzivasileiadi

Technology	Typical life-time years (cycles)	Power density Wkg⁻¹/kWm⁻³	Energy density Whkg⁻¹/kWhm⁻³	Typical discharge time	Recharge time	Response time	Operating temperature°C
Lead acid	3–15 (2000)	75–300/90–700	30–50/75	s–3 h	8–16 h	ms	25
NiCd	NiCd	150–300/75–700	45–80/<200	s–h	1 h	ms	–40 to45
Li-ion	8–15(4 × 10³)	230–340/1300–10000	100–250/250–620	min–h	Min–h	ms–s	–10 to50
NaS	12–20(>2000)	90–230/120–160	150–240/<400	s–h	9 h	ms	300
Na-NiCl	12–20(>1000)	130–160/250–270	125/150–200	min–h	6–8 h	ms	270 to 350
Zn-Br₂ FB	5–10(>2000)	50–150/1–25	60–80/20–35	s–10 h	4 h	<1ms	20 to 50
V-Redox FB	10–20(13 × 10³)	NA/0.5–2	75/20–35	s–10 h	Min	<1ms	0 to 40
Flywheel	>20 (107)	400–1600/5000	5–130/20–80	15 s–15min	<15min	ms–s	20 to 40
Super/DL capacitors	>20 (5 × 10⁵)	0.1–10/40000–120000	0.1–15/10–20	ms–1 h	s–min	ms	–40 to85
Pumped-Hydro	50–100(>500)	NA/0.1–0.2	0.5–1.5/0.2–2	h–days	Min–h	s–min	Ambient
Compressed air (CAES)	25–40 (No limit)	NA/0.2–0.6	30–60/12	h–days	Min–h	1–15min	Ambient

From a future technology deployment perspective, different energy storage technologies have a differing level of maturity (International Electrotechnical Commission, 2011). Some technologies are suitable for immediate deployment for grid applications whereas a number of others still require further research and development to improve performance and lifetime and also develop low cost mass production processes before these can be deployed on a large scale. Aside from the technical challenges described above, consideration also needs to be given to the economics and business models for the energy storage deployment. A simple methodology developed by the US DoE can be used to understand the monetary values of different technologies. The method involves taking into consideration the economic factors from location where the storage is to be located, the market, the asset type and who the owner is, then factoring in how the storage system will be used, the benefits and finally calculating a monetary value (Akhil et al., 2013). The methodology needs to be expanded to include modifications to also take account of additional devaluing items such as environmental and installations costs.

The key future requirements and challenges that energy storage technologies face are low installation costs, high durability and reliability, long service lifetimes and high round trip efficiency (U.S. Energy Information Administration, 2014). Furthermore, operation and maintenance costs are also critical in large scale deployment of energy storage solutions for the grid. Clearly, these requirements bring forward the need for scientific advances in the existing technologies which allow either a reduction in manufacturing/materials costs or longer service times etc. Many energy storage solutions which are commercially available have not been designed for large scale deployment, and this is holding these technologies back for grid deployment. Key advances in materials science or engineering as well as process science exist and provide ample opportunities for researchers in the future.

MEMBRANE SEPARATION TECHNOLOGIES

The developments in membranes for gas separation have much wider implication in low emission power generation, for controlling gas atmosphere and production of hydrogen and oxygen for a range of applications. In this regard a number of electrochemical gas separation technologies, mostly based on solid electrolytes are under development. All solid state electrochemical cells, where the electrolyte membrane is an oxygen-ion or a proton conductor (pure ionic or mixed ionic/electronic), can be used for selective transport of oxygen or hydrogen in the form of ionic flux thereby acting as electrochemical filters for molecular transport of oxygen or hydrogen.

Apart from the hydrogen production technologies discussed above, there has been a strong emphasis on developing both proton conducting polymer and oxygen-ion conducting ceramic membranes for high purity oxygen production for medical (e.g., home care oxygen therapy), defense, space and clean energy production applications (Badwal and Ciacchi, 2001; Badwal et al., 2003; Phair and Badwal, 2006a; Ursua et al., 2012). For example, in a concept described by Giddey et al. (Ursua et al., 2012), an electrolysis cell based on the proton conducting polymer membrane NAFION was used to split water to produce oxygen on one side of the cell with protons migrating through the membrane to the other electrode/electrolyte interface which then reacted with oxygen from air supplied to produce water. In this mode of operation, one half of the electrochemical cell operated in the water electrolysis mode and the other half in the fuel cell mode thus reducing by 30–40% the power required by a normal water electrolysis cell (Ursua et al., 2012).

The ceramic membranes for high purity oxygen production are based on O^{2-} conducting solid electrolyte such as zirconia, ceria, and bismuth oxide doped with divalent or trivalent cations such as Ca^{2+}, Y^{3+}, Yb^{3+}, Sc^{3+}, Gd^{3+}, etc. (Badwal and Ciacchi, 2000). Although solid electrolytic cells based on pure ionic conductors are useful for oxygen removal to generate inert atmospheres or for oxygen level control, their use for large scale oxygen production is limited to specific applications (Badwal et al., 2003) due to the large energy input (applied voltage) required to drive across the electrochemical cell. For bulk oxygen production applications such as oxyfuel combustion, mixed ionic/electronic conductors (MIEC) have been considered and technology developed based on such materials (Zhang et al., 2011). These devices typically rely on oxygen partial pressure differential across the MIEC membrane to transport oxygen through the membrane.

In hydrogen production from fossil fuels, hydrogen separation and purification is a key step. The HT ceramic based proton conducting membranes have been considered for pumping hydrogen across an electrochemical cell (Phair and Badwal, 2006b; Gallucci et al., 2013). The use of pure ionic conducting membranes is energy intensive as these devices are driven by external voltage or current. However, mixed proton/electronic conducting membranes are of interest for separating hydrogen for example from a mixture of CO_2 and H_2 following gasification of coal or reforming of NG. Recent reviews discuss many proton conducting membrane materials and gas separation reactors (Phair and Badwal, 2006b; Gallucci et al., 2013).

In the area of gas separation membranes, there are major technical challenges in fabrication of composite structures, chemical and thermal compatibility between components of the composite structure, interface coherency, optimization of the microstructure, lifetime issues in real operating environments (integrated into coal gasification, NG reforming plants), fabrication of support structures for deposition of thin films of the membrane material with optimal properties to achieve desired hydrogen or oxygen permeation rates and selectivity to the transporting specie. Some of the other major issues are related to fabrication, up-scaling and to have good mechanical strength and toughness as well as good chemical stability in real operating environments.

ELECTROCHEMICAL REACTORS FOR ENERGY CONVERSION AND STORAGE

Interest in electrochemical reactors stem from the fact that energy can be converted from one form to another more useful form for easy storage and transportation (for example, hydrogen, ammonia, or syn gas—a precursor for the liquid fuel production—with the use of a renewable energy source). In electrochemical cells, electrochemical processes can also be used to produce value added fuels or chemicals. Several different types of systems based on liquid and solid electrolytes have been proposed. The major advantage of the solid electrolyte systems is that both reactant and product chemicals are separated by the electrolyte membrane and a wide range of operating conditions are available to suit a particular chemical/electrochemical reaction. Two types of systems under development are based on oxygen-ion or proton conducting electrolytes. The selectivity to partial oxidation/reduction reaction can be controlled by the suitable choice of catalytic electrodes or catalyst/electrode mixtures, by the careful control over migration rates of oxygen-ion or protons and cell operating conditions. In the three sections below some electrochemical processes are briefly described.

Partial Oxidation, Hydrogenation, Dehydrogenation Reactors

In these reactors either a pure O^{2-} or H^+ conducting (IC) or a mixed O^{2-}/e^- or H^+/e^- conducting (MIEC) membrane is used to separate the reactant and products (Iwahara et al., 2004; Sundmacher et al., 2005; Wei et al., 2013). These materials have typically perovskite (ABO_3), fluorite (MO_2), or pyrochlore ($A_2B_2O_7$) structures. Often electrode/catalyst layer is applied to both sides of the ion conducting membrane as shown in Figure 14. The O^{2-} or H^+ ions migrate under an applied electric field or partial pressure differential of the migrating specie and either oxidize or reduce the reactant to produce fuels or value added chemicals.

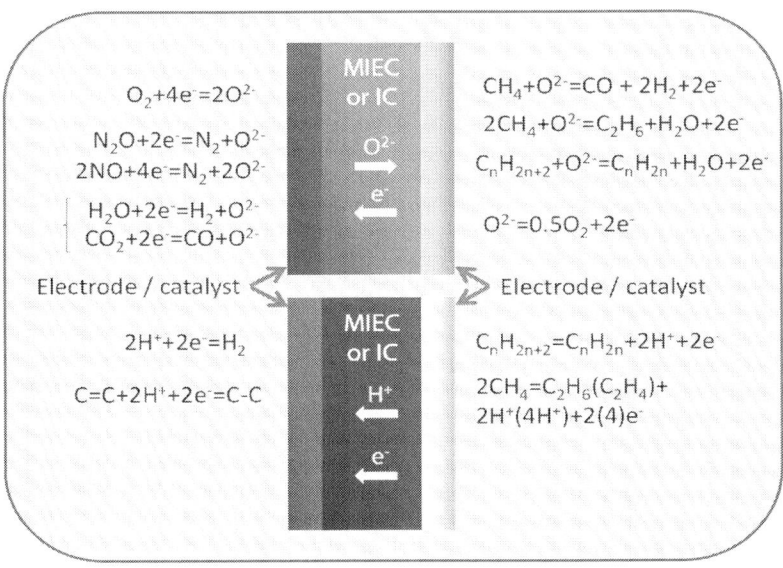

Figure 14: Basic operating principle of O^{2-} and H^+ electrochemical reactors for fuel and chemical production.

Dense ceramic membranes with mixed O^{2-}/e^- have been used for the conversion of methane to syngas, oxidative coupling of methane to higher hydrocarbons (C_2), conversion of ethylbenzene to styrene, and oxidative dehydrogenation of alkanes to olefins and conversion of pollutants such as N_2O and NO to N_2 by extracting oxygen (Wei et al., 2013). Similarly proton conducting membranes can be used for hydrogenation or dehydrogenation reaction by adding or stripping hydrogen from organic compounds (CH_4 to C_2H_4, C_2H_6; C–C to C=C; C=C to C–C) (Sundmacher et al., 2005; Wei et al., 2013) (Figure 14). Various processes, electrochemical reactors and materials of construction based on dense O^{2-} and H^+ conducting ceramic membranes have been reviewed extensively in the literature (Iwahara et al., 2004; Sundmacher et al., 2005; Wei et al., 2013).

There are a number of material, fabrication, design and up-scaling challenges for a given type of electrochemical reactor. Often materials are exposed to strongly oxidizing or reducing conditions at HTs. This chemical stability and thermal compatibility of all cell components needs to be addressed. The selectivity to a particular reaction and production rates often compete and for given reaction conditions undesirable products can easily form. Apart from the general criteria of high ionic flux for the transporting specie and thermal and chemical stability of the membrane materials, for the type of electrochemical reaction to take place, several materials and operating conditions need to be optimized.

Waste to Fuels and Value Added Products

The electrochemical conversion of waste products such as biomass (agricultural and forest residue), municipality, or industrial waste to value added chemicals and fuels is an area of enormous interest globally from the commercial as well as environmental view point. These waste materials can be converted to electricity, heat, gaseous (CO, H_2, CH_4), or liquid fuels (methanol, ethanol, biodiesel, etc.) by employing HT processes which are highly efficient and CO_2 neutral.

Microbial Electrochemical System for Hydrogen and Biofuel Production

One of the rapidly developing areas for conversion of waste to value added chemicals is based on a microbial electrochemical system called microbial electrolysis (Logan and Rabaey, 2012; Wang and Ren, 2013). In a microbial electrolysis cell (MEC), the organic and inorganic parts of the waste material in the anode chamber of the cell are oxidized with the help of microorganisms (electrochemically active bacteria) to CO_2 and electrons. The electrons are passed on to the electrode, and protons thus generated are transported through the electrolyte. In the cathode chamber, the protons can either react with electrons supplied from the external circuit to produce hydrogen (as a fuel) or can be made to react (hydrogenation) with another species to produce other value added chemicals such as biofuels.

Figure 15 illustrates this process schematically. The theoretical voltage required for producing hydrogen by MEC is 0.41 V compared to 1.2 V for conventional water electrolysis, however, applied voltages as high as 1 V is required for MEC to achieve practical hydrogen generation rates (Logan and Rabaey, 2012). By employing renewable and waste materials in MEC, the hydrogen production rates of more than three times have been achieved compared to those obtained by dark fermentation (Wang and Ren, 2013). The major challenge for commercialization of this technology is the cost of precious metal catalyst electrodes and other associated materials (Logan and Rabaey, 2012), and the sluggish reaction rates to achieve practical hydrogen or other chemical production rates.

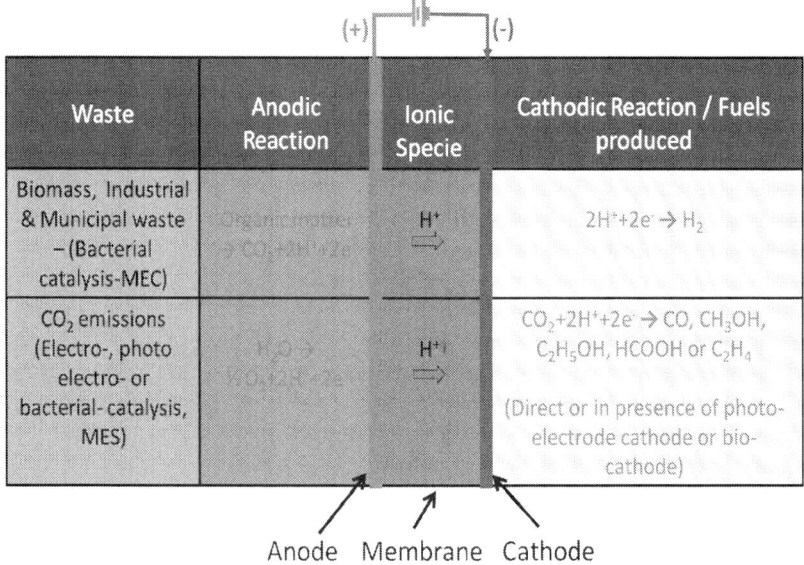

Figure 15: Electrochemical reactions involved in various processes for producing fuels and value-added chemicals from waste.

Conversion of Co_2 to Fuels with Renewable Energy

Another emerging area under development energy conversion and storage involves the utilization of CO_2 as the feedstock to electrochemically synthesize fuels and certain specialty chemicals such as carbon monoxide, methanol, formic acid, methane, ethylene, and oxalic acid (Jitaru, 2007). The utilization of electricity from renewable sources to convert CO_2 to high energy density fuels can help in alleviating the challenges of intermittent nature of the renewable sources by storing energy in the form of high energy density fuels, as well as addressing the liquid fuel shortage for the transport sector. Apart from the production of fuels, some products formed by CO_2 conversion may also be suitable as a feedstock for the chemical, pharmaceutical, and polymer industries. A number of review articles provide details on the methods of CO_2 reduction, electrode/electrolyte systems under consideration, various chemical products that can be produced and current status of the technology (Jitaru, 2007; Hori, 2008; Lee et al., 2009; Beck et al., 2010; Li, 2010; Whipple and Kenis, 2010; Hu et al., 2013; Jhong et al., 2013; Qiao et al., 2014). The processes employed for the electrochemically conversion of CO_2 include electro-catalysis (direct electrochemical conversion), photo electro-catalysis and bacteria-assisted electro-catalysis as shown schematically in Figures 14, 15. Although many processes are at an early stage of technological developments and there are concerns about the economic viability, these processes are discussed briefly in the following sections.

Direct Electrochemical Conversion

The main electrolyte systems under consideration for the direct electrochemical conversion of CO_2 are divided into low or ambient temperature systems [aqueous, non-aqueous (Cook et al., 1990; Hara et al., 1995; Hara and Sakata, 1997; Jitaru, 2007;Ogura, 2013) and PEM-based (Delacourt et al., 2008; Aeshala, 2013) electrolytes]; and HT systems [molten carbonate (Licht et al., 2010) and solid oxide (Stoots, 2006, 2010; Bidrawn et al., 2008;Hartvigsen et al., 2008; Ebbesen and Mogensen, 2009; Zhan et al., 2009; Fu et al., 2011; Graves et al., 2011; Narasimhaiah and Janardhanan, 2013) electrolytes—in the 700–1000°C range]. In the direct electro-catalysis process, CO_2 is supplied as a feedstock to the cathode chamber of the cell for reduction. In case of LT electrolyte systems (aqueous and PEM electrolytes), water is supplied to the anode as a source of protons for reaction at the cathode (Delacourt et al., 2008; Aeshala, 2013;Ogura, 2013). The protons transported through the electrolyte to the cathode are made to react with CO_2 to produce fuels or chemicals (Figures 14, 15). The competing reaction in aqueous- and PEM-based electrolytes is the hydrogen evolution that should be avoided, otherwise it results in wastage of energy input to the process if hydrogen is not the required chemical. Most metallic electrodes employed in the process yield CO and HCOOH, however, copper can also yield hydrocarbons such as methane and ethylene (Jitaru, 2007). Ogura has recently reported the CO_2 reduction to ethylene on copper halide confined copper mesh electrode with current efficiency of up to 80% and selectivity of up to 87% (Ogura, 2013).

In a molten carbonate electrolyte system, CO_2 is dissolved in the carbonate bath and is reduced to CO via the electrolysis process. The electrical energy input for the endothermic CO_2 reduction reaction reduces as the process is carried out at HTs with solar thermal energy input (Licht et al., 2010). In a solid oxide electrolyte system, CO_2 supplied to the cathode is reduced to CO and oxygen anions thus formed are transported through the solid electrolyte to produce oxygen at the anode. The solid oxide electrolyte cells have also been investigated for co-electrolysis of CO_2 and water (Figure 14). In this case, steam and CO_2 are both supplied to the cathode that results in formation of syn gas (H_2 + CO) at the cathode and oxygen at the anode (Stoots, 2006; Hartvigsen et al., 2008). Although the electrochemical conversion of CO_2 to different hydrocarbon fuels has been demonstrated by a number of investigators, the real challenges are to improve the conversion rates (CO_2 being a stable molecule and is difficult to reduce) and energy efficiencies to make the process commercially viable. Thus new catalysts, processes and materials need to be developed to reduce cell voltage losses and improve the selectivity and conversion efficiency (Whipple and Kenis, 2010; Hu et al., 2013). In a recent article, Jhong et al. have covered the current status, challenges, and future opportunities for electrochemical conversion of CO_2 to useful chemicals (Jhong et al., 2013).

Photo Electrochemical Conversion

In a photo electro-catalysis process, a photo-reduction electrode that consists of a semiconductor and a photo-catalyst is used as a cathode (Hu et al., 2013). The photons from the solar radiation, absorbed by the semiconductor cause the excited electrons transfer from valence to conduction band, that results in transfer of electrons to photo-catalysts. This electron transfer assists in the CO_2 reduction reaction involving protons transported through the electrolyte to produce CO and other organic compounds (Figure 15). It has been reported that the onset voltages for the CO_2 reduction process are significantly reduced by employing photo electrodes (cathode) compared to metallic electrodes (Kumar et al., 2012; Hu et al., 2013). Both aqueous and non-aqueous systems have been explored for the photo electrochemical reduction of CO_2. Higher solubility of CO_2 in non-aqueous electrolytes compared to aqueous electrolytes is favorable to achieve high current densities and increase selectivity over hydrogen evolution, however, other means such as high pressure and employing gas diffusion electrodes can be used for both types of electrolytes to increase CO_2 concentration. Some of the electrode/electrolyte systems investigated for the CO_2 reduction in aqueous media are p-Si / $NaSO_4$, p-CdTe and p-InP/tetraalkylammonium, p-GaAs in KCl, $HClO_4$, or Na_2CO_3 electrolyte (Barton et al., 2008; Kumar et al., 2012). Other photo electrodes explored for CO_2 reduction are Cu, Ag or Au, Pd nano particles attached to p-Si or p-InP (Barton et al., 2008; Kumar et al., 2012). Although the photo electrodes investigated for the non-aqueous electrolytes have been same as for aqueous electrolytes, the popular electrolyte used has been methanol, due to its high CO_2 solubility. The chemicals produced, and the Faradaic efficiency and selectivity of the chemical produced depends on the photo electrode and the supporting electrolyte used. These systems have been reviewed quite extensively by Kumar et al. (2012) and more details on performance of these systems can be found in this review. The low efficiencies and current densities achieved, and the high costs of the catalysts used in this process are still some of the major challenges for this technology.

Bacterial-Assisted Electrochemical Conversion

In bacteria-assisted electrosynthesis, the microorganisms at the cathode of the electrochemical cell assist in the reduction of CO_2 to fuels or value added chemicals. This process is also called microbial electrosynthesis (MES) (Wang and Ren, 2013). As depicted in Figure 15, the process involves protons transported through the electrolyte, electrons delivered to cathode and CO_2 supplied to the cathode camber. It is claimed that with electric input from renewable energy sources, the microbes can harvest the solar energy at 100 times the efficiency of a biomass-based fuel/chemical production (Wang and Ren, 2013). The formation of products that have already been demonstrated from this route by employing various types of cultures, are methane, acetate, and oxo-butyrate. In another variation to the MEC, described in Section Microbial Electrochemical System for Hydrogen and Biofuel Production, if the protons transported through the electrolyte to cathode (biocathode) are made to react with the CO_2, other chemicals can be formed in preference to hydrogen generation. In a recent study employing a MEC based on a cation exchange membrane, CO_2 was successfully converted to methane for a period of 188 days with an overall energy efficiency of 3.1% (Van Eerten-Jansen et al., 2011). The rates and quantities of the chemical produced by microbial synthesis and electrolysis cells, and the overall energy efficiencies are very low, and would require significant improvements to the synthesis process as well as the cell configuration to lower resistive losses in the various cell components for a large scale operation (Van Eerten-Jansen et al., 2011; Logan and Rabaey, 2012).

ELECTROCHEMICAL PROCESSES FOR AMMONIA PRODUCTION

Ammonia is an excellent energy storage media with infrastructure for its transportation and distribution already in place in many countries. Liquid ammonia has a hydrogen content of 17.6 wt% and therefore can be utilized as a source of hydrogen at distributed sites. By comparison the hydrogen content in methanol is only 12.5 wt%. Over 200 million metric tons of ammonia is produced per annum globally and in terms of production volumes, it is one of the major chemicals produced. Current ammonia production processes are highly energy intensive (Giddey et al., 2013). Ammonia is the intermediate chemical for the production of many chemicals including over 80% utilization for fertilizer production with other important uses including the manufacture of explosives, pharmaceutical chemicals and other industrial processes such as synthesis of specialty ceramic powders and refrigeration.

Ammonia is produced at present through the well-known Haber-Bosch process. Given the high energy consumption and high capital cost of the process requiring hydrogen and nitrogen to react on an iron-based catalyst at HTs (up to 550°C) and high pressures (up to 300 bar). In view of this a number of alternative processes are under investigation. Amongst many approaches, electrochemical routes have the potential to produce ammonia under very mild conditions of temperature and pressure and at a lower cost compared with the Haber-Bosch process of ammonia production (Giddey et al., 2013).

The various electrochemical routes for ammonia production are differentiated by the type of electrolyte used and the operating temperature regime. These can be broadly divided into four categories: (a) liquid electrolytes operating near room temperature,

(b) molten salt electrolytes operating in the 300–500°C range, (c) composite electrolytes consisting of a solid and a low melting molten salt, and (d) solid electrolytes with operating temperature range from room temperature (typically polymers) to 800°C (ceramics). Other materials of construction are based on the type of system selected. Typical operation of an electrochemical ammonia production process is described in Figure 16. These systems have been discussed in detail in recent reviewed articles (Amar et al., 2011;Giddey et al., 2013; Garagounis et al., 2014) and involve supply of hydrogen at the anode/electrolyte interface, migration of protons through the electrolyte and reaction with N_2 over a cathode catalyst to form ammonia. Materials requirements include high ionic conductivity in the electrolyte and chemical stability under operating conditions and thermo-mechanical compatibility between various cell components. The catalyst on the nitrogen side plays a critical role.

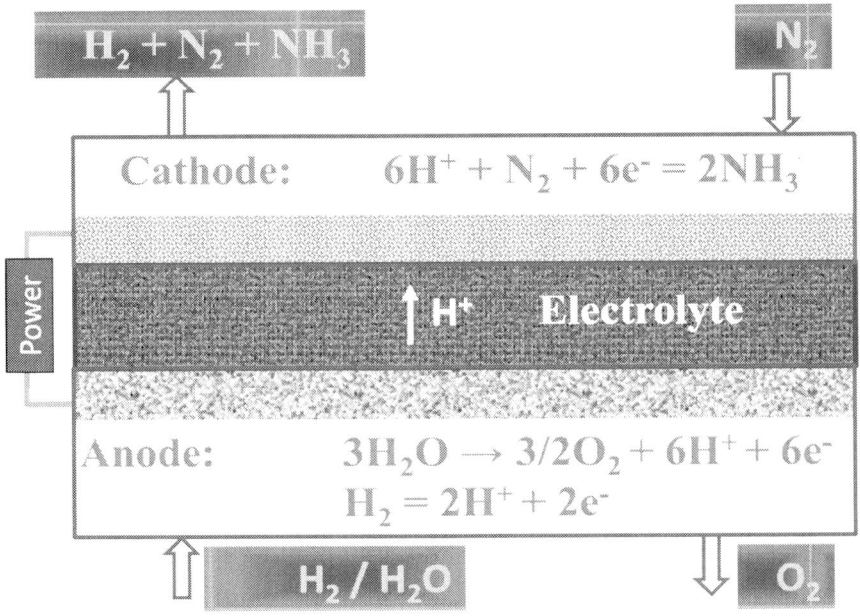

$$H_2 + N_2 + NH_3 \qquad N_2$$

Cathode: $6H^+ + N_2 + 6e^- = 2NH_3$

$$\uparrow H^+ \quad \text{Electrolyte}$$

Anode: $3H_2O \rightarrow 3/2O_2 + 6H^+ + 6e^-$

$$H_2 = 2H^+ + 2e^-$$

Power

$$H_2 / H_2O \qquad O_2$$

Figure 16: The operating principle of ammonia production in a solid state electrochemical cell.

Two critical performance parameters that determine the overall process efficiency are the current efficiency and ammonia production rates. The current efficiency or conversion rates determine the percentage of protons flowing through the electrolyte that are effectively utilized in ammonia formation. The ammonia production rates are defined in number of moles of ammonia produced per unit cell area per unit time typically expressed as $mol.cm^{-2}.s^{-1}$. Both high ammonia production rates and high current efficiency are essential for the economic viability of the process. The higher operating temperature improves kinetics of reaction between nitrogen and hydrogen and would allow integration with thermal solar or nuclear power plants for heat input. However, the thermodynamics of the reaction favors operation at LTs and high pressures and hence offer the potential to use low cost materials.

This technology is at an early stage of development requiring considerable work on the development of cell materials and ammonia production catalyst. The ammonia production rates achieved by various electrochemical processes are in the $10^{-13-10-8}$ $mol.cm^{-2}.s^{-1}$ and are too low for the process to be economically viable. The highest production rate reported was for a PEM-based electrochemical reactor. Often high production rates are quoted at low current densities and for high hydrogen conversion rates (over 50%) reported in the literature, the ammonia production rates are low (Giddey et al., 2013). At least another order of magnitude increase in ammonia production rates with conversion efficiency well above 50% at current densities above 0.25 $A.cm^{-2}$ would make the process technically feasible for consideration of the technology for commercialization. Lifetime, degradation rates, cost of materials and fabrication processes, and up-scaling are some of the other considerations.

CONCLUSIONS

Electrochemical energy technologies are already contributing substantially to reduction of pollution and greenhouse gas emissions, in process control and via increasing energy conversion efficiency. The growing demand for technologies that can stabilize power generation and delivery is driving research toward developing new technologies. This is increasing the number of systems under investigation across the entire innovation chain from very early stage research through to development of conventional devices to increase performance and reduce cost. As with all new technologies there remain many technical challenges facing the developers of future electrochemical power systems, however, the increased understanding of the value of these technologies is leading to an increase in the scale of programs looking to improve these technologies. It is unclear which new technologies will emerge as leaders in the future power market but it is clear that there will be significant improvement over current devices in terms of cost reduction, performance, and availability over the next decade. This will go beyond lone new electrochemical cell chemistries and will increasingly involve the development of highly integrated hybrid systems that take advantage of the strengths of multiple technology features.

ACKNOWLEDGMENT

The authors would like to thank Dr. Ani Kulkarni for reviewing this manuscript.

REFERENCES

1. Abraham, K. M., and Jiang, Z. (1996). Preparation and electrochemical characterization of micron-sized spinel $LiMn_2O_4$. *J. Electrochem. Soc.* 143, 1–5. doi: 10.1149/1.1836378

2. Aeshala, L. M. (2013). Effect of cationic and anionic solid polymer electrolyte on direct electrochemical reduction of gaseous CO_2 to fuel. *J. CO_2 Util.* 3–4, 49–55. doi: 10.1016/j.jcou.2013.09.004

3.   Akhil, A. A., Huff, G., Currier, A. B., Kaun, B. C., Rastler, D. M., Chen, S. B., et al. (2013). *DOE/EPRI 2013 Electricity Storage Handbook in Collaboration with NRECA*. Albuquerque, NM: Sandia National Laboratories, SAND2013–5131.

4.   Alexander, B. R., Mitchell, R. E., and Gür, T. M. (2011). Steam-carbon fuel cell concept for cogeneration of hydrogen and electrical power. *J. Electrochem. Soc.* 158, B505–B513. doi: 10.1149/1.3560475

5.   Amar, I. A., Lan, R., Petit, C. T. G., and Tao, S. (2011). Solid-state electrochemical synthesis of ammonia: a review. *J. Solid State Electrochem.* 15, 1845–1860. doi: 10.1007/s10008-011-1376-x

6.   Arriaga, L. G., Martínez, W., Cano, U., and Blud, H. (2007). Direct coupling of a solar-hydrogen system in Mexico. *Int. J. Hydrogen Energy* 32, 2247–2252. doi: 10.1016/j.ijhydene.2006.10.067

7.   Aurbach, D., Pollak, E., Elazari, R., Salitra, G., Kelley, C. S., and Affinito, J. (2009). On the surface chemical aspects of very high energy density, rechargeable Li-sulfur batteries. *J. Electrochem. Soc.* 156 A694–A702. doi: 10.1149/1.3148721

8.   Badwal, S. P. S. (1994). "Ceramic superionic conductors" in *Materials Science and Technology, A Comprehensive Treatment*, Vol. 11, eds R. W. Cahn, P. Haasen, and E. J. Kramer (Vol. ed M.V. Swain) (Weinheim: VCH Verlagsgesellschaft), 567–633.

9.   Badwal, S. P. S., and Ciacchi, F. T. (2000). Oxygen-ion conducting electrolyte materials for solid oxide fuel cells. *Ionics* 6, 1–21. doi: 10.1007/BF02375543

10. Badwal, S. P. S., and Ciacchi, F. T. (2001). Ceramic membrane technologies for oxygen separation. *Adv. Mater.* 13, 993–996. doi: 10.1002/1521-4095(200107)13:12/13<993::AID-ADMA993>3.0.CO;2-#

11. Badwal, S. P. S., Ciacchi, F. T., Zelizko, V., and Giampietro, K. (2003). Oxygen removal and level control with Zirconia - yttria membrane cells. *Ionics* 9, 315–320. doi: 10.1007/

BF02376580

12. Badwal, S. P. S., Giddey, S., Kulkarni, A., and Munnings, C. (2014). Review of progress in high temperature solid oxide fuel cells. *J. Aust. Cer. Soc.* 50, 23–37. Available online at: http://www.austceram.com/JAC-2014-1/Article-PDFs/3_JACS%2050_Badwal_23-37.pdf

13. Badwal, S. P. S., Giddey, S., and Munnings, C. (2013). Hydrogen production via solid electrolyte routes. *WIREs Energy Environ.* 2, 473–487. doi: 10.1002/wene.50

14. Bae, C. H. (2001). *Cell Design and Electrolytes of a Novel Redox Flow Battery.* Ph.D. Thesis, University of Manchester (UMIST), Manchester, UK.

15. Bae, C. H., Roberts, E. P. L., Chakrabarti, M. H., and Saleem, M. (2011). All-chromium redox flow battery for renewable energy storage. *Int. J. Green Energy* 8, 248–264. doi: 10.1080/15435075.2010.549598

16. Barton, E. E., Rampulla, D. M., and Bocarsly, A. B. (2008). Selective solar-driven reduction of CO_2 to methanol using a catalysed p-GaP based photoelectrochemical cell. *J. Am. Chem. Soc.* 130, 6342–6344. doi: 10.1021/ja0776327

17. Beck, J., Johnson, R., and Naya, T. (2010). *Electrochemical Conversion of Carbon Dioxide to Hydrocarbon Fuels.* EME 580 Spring, 1–42. Available online at:http://www.ems.psu.edu/~elsworth/courses/egee580/2010/Final%20Reports/co2_electrochem.pdf (Accessed September 2, 2014).

18. Bidrawn, F., Kim, G., Corre, G., Irvine, J. T. S., Vohs, J. M., and Gorte, R. J. (2008). Efficient reduction of CO_2 in a solid oxide electrolyser. *Electrochem. Solid State Lett.* 11, B167–B170. doi: 10.1149/1.2943664

19. Braff, W. A., Bazant, M. Z., and Buie, C. R. (2013). Membrane-less hydrogen bromine flow battery. *Nat. Commun.* 4, 2346. doi: 10.1038/ncomms3346

20. Brisse, A., Schefold, J., and Zahid, M. (2008). High temperature water electrolysis in solid oxide cells. *Int. J. Hydrogen Energy*

33, 5375–5382. doi: 10.1016/j.ijhydene.2008.07.120

21. Brown, L. F. (2001). A comparative study of fuels for on-board hydrogen production for fuel-cell-powered automobiles. *Int. J. Hydrogen Energy.* 26, 381–397. doi: 10.1016/S0360-3199(00)00092-6

22. Bruce, P. G., Freunberger, S. A., Hardwick, L. J., and Tarascon, J.-M. (2012). $Li-O_2$ and Li-S batteries with high energy storage. *Nat. Mater.* 11, 19–29. doi: 10.1038/nmat3191

23. Bruce, P. G., Hardwick, L. J., and Abraham, K. M. (2011). Lithium-air and lithium-sulfur batteries. *MRS Bull.* 36, 506–511. doi: 10.1557/mrs.2011.157

24. Burke, A. (2010). Ultracapacitor technologies and application in hybrid and electric vehicles. *Int. J. Energy Res.* 34, 133–151. doi: 10.1002/er.1654

25. Carrette, L., Friedrich, K. A., and Stimming, U. (2005). Fuel cells: principles, types, fuels, and applications. *Chemphyschem*1, 162–193. doi: 10.1002/1439-7641(20001215)1:4<162::AID-CPHC162>3.0.CO;2-Z

26. Carter, D., and Wing, J. (2013). *The Fuel Cell Today Industry Review (2013)*. Fuel Cell Today. Available online at:http://www.fuelcelltoday.com/media/1889744/fct_review_2013.pdf (Accessed September 2, 2014).

27. Catherino, H. A., Feres, F. F., and Trinidad, F. (2004). Sulfation in lead–acid batteries. *J. Power Sources* 129, 113–120. doi: 10.1016/j.jpowsour.2003.11.003

28. Chakrabarti, M. H., Dryfe, R. A. W., and Roberts, E. P. L. (2007). Evaluation of electrolytes for redox flow battery applications. *Electrochim. Acta* 52, 2189–2195. doi: 10.1016/j.electacta.2006.08.052

29. Chatzivasileiadi, A., Ampatzi, E., and Knight, I. (2013). Characteristics of electrical energy storage technologies and their applications in buildings. *Renew. Sustain. Energy Rev.* 25, 814–830. doi: 10.1016/j.rser.2013.05.023

30. Cheng, J., Zhang, L., Yang, Y., Wen, Y., Cao, G., and Wang, X. (2007). Preliminary study of single flow zinc-nickel battery. *Electrochem. Commun.* 9, 2639–2642. doi: 10.1016/j. elecom.2007.08.016

31. Chieng, S. C., and Skyllas-Kazacos, M. (1992). Modification of Daramic, microporous separator, for redox flow battery applications. *J. Membr. Sci.* 75, 81–91. doi: 10.1016/0376-7388(92)80008-8

32. Choudhary, T. V., Sivadinarayana, C., and Goodman, D. W. C. (2001). Catalytic ammonia decomposition: COx-free hydrogen production for fuel cell applications. *Catal. Lett.* 72, 197–201. doi: 10.1023/A:1009023825549

33. Clarke, R. E., Giddey, S., Ciacchi, F. T., Badwal, S. P. S., Paul, B., and Andrews, J. (2009). Direct coupling of an electrolyser to a solar PV system for generating hydrogen. *Int. J. Hydrogen Energy* 34, 2531–2542. doi: 10.1016/j.ijhydene.2009.01.053

34. Clarke, R. L., Dougherty, B., Mohanta, S., and Harrison, S. (2004). "Abstract 520, Cerium-Zinc regenerative fuel cell," in *Joint International Meeting: 206th Meeting of the Electrochemical Society/2004 Fall Meeting of the Electrochemical Society of Japan* (Honolulu).

35. Cole, T. (1983). Thermoelectric energy conversion with solid electrolytes. *Science* 221, 915–920. doi: 10.1126/science.221.4614.915

36. Collins, J., Li, X., Pletcher, D., Tangirala, R., Stratton-Campbell, D., Walsh, F., et al. (2010). A novel flow battery: a lead acid battery based on an electrolyte with soluble lead(II). Part IX: electrode and electrolyte conditioning with hydrogen peroxide. *J. Power Sources* 195, 2975–2978. doi: 10.1016/j. jpowsour.2009.10.109

37. Cook, R. L., MacDuff, R. C., and Sammells, A. F. (1990). High rate gas phase CO_2 reduction to ethylene and methane using gas diffusion electrodes. *J. Electrochem. Soc.* 137, 607–608. doi: 10.1149/1.2086515

38. Cooper, A. (2004). Development of a lead acid battery for

hybrid electric vehicle. *J. Power sources* 133, 116–125. doi: 10.1016/j.jpowsour.2003.11.069

39. Coughlin, R. W., and Farooque, M. (1982). Thermodynamic, kinetic, and mass balance aspects of coal-depolarized water electrolysis. *Ind. Eng. Chem. Process Des. Dev.* 21, 559–564. doi: 10.1021/i200019a004

40. Damian, A., and Irvine, J. T. S. (2012). Development of tubular hybrid direct carbon fuel cell. *Int. J. Hydrogen Energy* 37, 19337–19344. doi: 10.1016/j.ijhydene.2012.02.104

41. Da Mota, N., Finkelstein, D. A., Kirtland, J. D., Rodriguez, C. A., Stroock, A. D., and Abruña, H. D. (2012). Membraneless, room-temperature, direct borohydride/cerium fuel cell with power density of over 0.25 W/cm^2. *J. Am. Chem. Soc.* 134, 6076–6079. doi: 10.1021/ja211751k

42. Delacourt, C., Ridgway, P. L., Kerr, J. B., and Newman, J. (2008). Design of an electrochemical cell making syngas (CO + H_2) from CO_2 and H_2O reduction at room temperature. *J. Electrochem. Soc.* 155, B42–B49. doi: 10.1149/1.2801871

43. Devanathan, R. (2008). Recent developments in proton exchange membranes for fuel cells. *Energy Environ. Sci.* 1, 101–119. doi: 10.1039/b808149m

44. Dong, L., Chen, Z., Yang, D., and Lu, H. (2013). Hierarchically structured graphene-based supercapacitor electrodes. *RSC Adv.* 3, 21183–21191. doi: 10.1039/c3ra44357d

45. Ebbesen, S. D., and Mogensen, M. (2009). Electrolysis of carbon dioxide in solid oxide electrolysis cells. *J. Power Sources*193, 349–358. doi: 10.1016/j.jpowsour.2009.02.093

46. Edwards, J. H., Badwal, S. P. S., Duffy, G., Lasich, J., and Ganakas, G. (2002). The application of solid state ionics technology for novel methods of energy generation and supply. *Solid State Ionics* 152–153, 843–852. doi: 10.1016/S0167-2738(02)00384-3

47. El-Genk, M. S., and Tournier, J. M. (2004). AMTEC/TE static converters for high energy utilization, small nuclear power

plants. *Energy Convers. Manag.* 45, 511–535. doi: 10.1016/S0196-8904(03)00159-6

48. Ellis, B. L., and Nazar, L. F. (2012). Sodium and sodium-ion energy storage batteries. *Curr. Opin. Solid State Mater. Sci.* 16, 168–177. doi: 10.1016/j.cossms.2012.04.002

49. European Commission. (2013). *The Future Role and Challenges of Energy Storage.* DG ENER Working Paper. Available online at: http://ec.europa.eu/energy/infrastructure/doc/energy-storage/2013/energy_storage.pdf (Accessed September 2, 2014).

50. Eustace, D. J. (1980). Bromine complexation in zinc-bromine circulating batteries. *J. Electrochem. Soc.* 127, 528–532. doi: 10.1149/1.2129706

51. Ewan, B. C. R., and Adeniyi, O. D. (2013). A demonstration of carbon-assisted water electrolysis. *Energies* 6, 1657–1668. doi: 10.3390/en6031657

52. Fang, B., Iwasa, S., Wei, Y., Arai, T., and Kumagai, M. (2002). A study of the Ce(III)/Ce(IV) redox couple for redox flow battery application. *Electrochim. Acta* 47, 3971. doi: 10.1016/S0013-4686(02)00370-5

53. Fisher, R. A., Watt, M. R., and Ready, W. J. (2013). Functionalized carbon nanotube supercapacitor electrodes: a review on pseudocapacitive materials. *ECS J. Solid State Sci. Technol.* 2, M3170–M3177. doi: 10.1149/2.017310jss

54. Fu, Q., Dailly, J., Brisse, A., and Zahid, M. (2011). High-temperature CO_2 and H_2O electrolysis with an electrolyte-supported solid oxide cell. *ECS Trans.* 35, 2949–2956. doi: 10.1149/1.3570294

55. Fujiwara, S., Kasai, S., Yamauchi, H., Yamada, K., Makino, S., Matsunaga, K., et al. (2008). Hydrogen production by high temperature electrolysis with nuclear reactor. *Prog. Nuclear Energy.* 50, 422–426. doi: 10.1016/j.pnucene.2007.11.025

56. Gallucci, F., Fernandez, E., Corengia, P., and van Sint Annal, M. (2013). Recent advances on membranes and membrane reactors for hydrogen production. *Chem. Eng. Sci.* 92, 40–66.

doi: 10.1016/j.ces.2013.01.008

57. Garagounis, I., Kyriakou, V., Skodra, A., Vasileiou, E., and Stoukides, M. (2014). Electrochemical synthesis of ammonia in solid electrolyte cells. *Front. Energy Res.* 2:1. doi: 10.3389/fenrg.2014.00001

58. García-Valverde, R., Espinosa, N., and Urbina, A. (2011). Optimized method for photovoltaic-water electrolyser direct coupling. *Int. J. Hydrogen Energy* 36, 10574–10586. doi: 10.1016/j.ijhydene.2011.05.179

59. Giddey, S., Badwal, S. P. S., and Kulkarni, A. (2013). Review of electrochemical ammonia production technologies and materials. *Int. J. Hydrogen Energy* 38, 14576–14594. doi: 10.1016/j.ijhydene.2013.09.054

60. Giddey, S., Badwal, S. P. S., Kulkarni, A., and Munnings, C. (2012). A comprehensive review of direct carbon fuel cell technology. *Progress Energy Combust. Sci.* 38, 360–399. doi: 10.1016/j.pecs.2012.01.003

61. Giddey, S., Ciacchi, F. T., and Badwal, S. P. S. (2010). High purity oxygen production with a polymer electrolyte membrane electrolyser. *J. Membr. Sci.* 346, 227–232. doi: 10.1016/j.memsci.2009.09.042

62. Girishkumar, G., McCloskey, B., Luntz, A. C., Swanson, S., and Wilcke, W. (2010). Lithium-air battery: promise and challenges. *J. Phys. Chem. Lett.* 1, 2193–2203. doi: 10.1021/jz1005384

63. Graves, C., Ebbesen, S. D., and Mogensen, M. (2011). Co-electrolysis of CO_2 and H_2O in solid oxide cells: performance and durability. *Solid State Ionics* 192, 398–403. doi: 10.1016/j.ssi.2010.06.014

64. Guth, U., Vonau, W., and Zosel, J. (2009). Recent developments in electrochemical sensor application and technology - a review. *Meas. Sci. Technol.* 20, 1–14. doi: 10.1088/0957-0233/20/4/042002

65. Hara, K., Kudo, A., and Sakata, T. (1995). High efficiency

electrochemical reduction of carbon dioxide under high pressure on a gas diffusion electrode containing Pt catalysts. *J. Electrochem. Soc.* 142, L57–L59. doi: 10.1149/1.2044182

66. Hara, K., and Sakata, T. (1997). Electrocatalytic formation of CH_4 from CO_2 on a Pt gas diffusion electrode. *J. Electrochem. Soc.* 144, 539–545. doi: 10.1149/1.1837445

67. Harrison, K., and Peters, M. (2013). "Renewable electrolysis integrated system development and testing," in *2013 DOE Hydrogen and Fuel Cells Program Review, National Renewable Energy Laboratory, Project ID: PD031* (Denver, CO).

68. Harrison, K. W., Martin, G. D., Ramsden, T. G., and Kramer, W. E. (2009). "The wind-to-hydrogen project: operational experience, performance testing, and systems integration," in *National Renewable Energy Laboratory. Technical Report NREL/TP-550-44082* (Denver, CO).

69. Harrop, P., Zhitomirsky, V., and Gonzalez, F. (2014). *Electrochemical Double Layer Capacitors: Supercapacitors 2014-2024*. IDTechEx. Available online at: http://www.idtechex.com/research/reports/electrochemical-double-layer-capacitors-supercapacitors-2014-2024-000378.asp (Accessed September 2, 2014).

70. Hartmann, P., Bender, C. L., Sann, J., Dürr, A. K., Jansen, M., Janek, J., et al. (2013). A comprehensive study on the cell chemistry of the sodium superoxide (NaO_2) battery. *Phys. Chem. Chem. Phys.* 15, 11661–72. doi: 10.1039/c3cp50930c

71. Hartvigsen, J., Elangovan, S., Frost, L., Nickens, A., Stoots, C., O'Brien, J., et al. (2008). Carbon dioxide recycling by high temperature co-electrolysis and hydrocarbon synthesis. *ECS Trans.* 12, 625–637. doi: 10.1149/1.2921588

72. Hasegawa, K., Kimura, A., Yamamura, T., and Shiokawa, Y. (2005). Estimation of energy efficiency in neptunium redox flow batteries by the standard rate constants. *J. Phys. Chem. Solids* 66, 593–595. doi: 10.1016/j.jpcs.2004.07.018

73. Hazza, A., Pletcher, D., and Wills, R. (2005). A novel flow

battery-a lead acid battery based on an electrolyte with soluble lead(II) IV. The influence of additives. *J. Power Sources* 149, 103–111. doi: 10.1016/j.jpowsour.2005.01.049

74. Hesenov, A., Meryemoglu, B., and Icten, O. (2011). Electrolysis of coal slurries to produce hydrogen gas: effects of different factors on hydrogen yield. *Int. J. Hydrogen Energy* 36, 12249–12258. doi: 10.1016/j.ijhydene.2011.06.134

75. Hiroko, K., Negishi, A., Nozaki, K., Sato, K., and Nakajima, M. (1994). *Redox Battery*. U.S. Patent US5318865.

76. Hiroko, K., Negishi, A., Nozaki, K., Sato, K., and Nakajima, M. (1997). *Redox Battery*. European Patent EP0517217 A1.

77. Hori, Y. (2008). "Electrochemical CO_2 reduction on metal electrodes," in *Modern Aspects of Electrochemistry*, eds C. G. Vayenas, R. E. White, and M. E. Gamboa-Aldeco (New York, NY: Springer), 89–189. doi: 10.1007/978-0-387-49489-0_3

78. Howlett, P. C., MacFarlane, D. R., and Hollenkamp, A. F. (2004). High lithium metal cycling efficiency in a room-temperature ionic liquid. *Electrochem. Solid State Lett.* 7, A97–A101. doi: 10.1149/1.1664051

79. Hu, B., Guild, C., and Suib, S. L. (2013). Thermal, electrochemical and photochemical conversion of CO_2 to fuels and value-added products. *J. CO_2 Util.* 1, 18–27. doi: 10.1016/j.jcou.2013.03.004

80. Imanishi, N., and Yamamoto, O. (2014). Rechargeable lithium-air batteries: characteristics and prospects. *Mater. Today* 17, 24–30. doi: 10.1016/j.mattod.2013.12.004

81. International Electrotechnical Commission. (2011). *Electrical Energy Storage White Paper*. Available online at:http://www.iec.ch/whitepaper/pdf/iecWP-energystorage-LR-en.pdf

82. International Energy Statistics. (2011). *U.S. Energy Information Administration*. Available online at:http://www.eia.gov/cfapps/ipdbproject/iedindex3.cfm?tid=90&pid=44&aid=8 (Accessed September 2, 2014).

83. IPHE. (2012). "2012 hydrogen and fuel cell global

commercialization and development update," in *International Partnership for Hydrogen and Fuel Cells in the Economy (IPHE)*. Available online at:http://www.iphe.net/docs/Commercialization_Rpt_Final_070913.pdf (Accessed September 2, 2014).

84. Iwahara, H., Asakura, Y., Katahira, K., and Tanaka, M. (2004). Prospect of hydrogen technology using proton-conducting ceramics. *Solid State Ionics* 168, 299–310. doi: 10.1016/j.ssi.2003.03.001

85. Jayakumar, A., Javadekar, A., Gissinger, J., Vohs, J. M., Huber, G. W., and Gorte, R. J. (2013). The stability of direct carbon fuel cells with molten Sb and Sb-Bi alloy anodes. *AICHE J.* 59, 3342–3348. doi: 10.1002/aic.13965

86. Jhong, H. R. M., Ma, S., and Kenis, P. J. A. (2013). Electrochemical conversion of CO_2 to useful chemicals: current status, remaining challenges, and future opportunities. *Curr. Opin. Chem. Eng.* 2, 191–199. doi: 10.1016/j.coche.2013.03.005

87. Jia, C., Liu, J., and Yan, C. (2010). A significantly improved membrane for vanadium redox flow battery. *J. Power Sources* 195, 4380–4383. doi: 10.1016/j.jpowsour.2010.02.008

88. Jin, X., and Botte, G. G. (2010). Understanding the kinetics of coal electrolysis at intermediate temperatures. *J. Power Sources* 195, 4935–4942. doi: 10.1016/j.jpowsour.2010.02.007

89. Jitaru, M. (2007). Electrochemical carbon dioxide reduction-fundamental and applied topics (Review). *J. Univ. Chem. Technol. Metal.* 42, 333–344.

90. Google Scholar

91. Kamaya, N., Homma, K., Yamakawa, Y., Hirayama, M., Kanno, R., Yonemura, M., et al. (2011). A lithium superionic conductor. *Nat. Mater.* 10682–10686. doi: 10.1038/nmat3066

92. Kang, S. Y., Mo, Y., Ong, S. P., and Ceder, G. (2014). Nanoscale stabilization of sodium oxides: implications for $Na\text{-}O_2$ batteries. *Nano Lett.* 14, 1016–1020. doi: 10.1021/

nl404557w

93. Knight, C., Cavanagh, K., Munnings, C., Moore, T., Cheng, K. Y., and Kaksonen, A. H. (2013). "Application of microbial fuel cells to power sensor networks for ecological monitoring," in *Wireless Sensor Networks and Ecological Monitoring*, eds S. C. Mukhopadhyay and J. A. Jiang (Berlin; Heidelberg: Springer), 151–178. doi: 10.1007/978-3-642-36365-8_6

94. Kraytsberg, A., and Ein-Eli, Y. (2011). Review on Li–air batteries—opportunities, limitations and perspective. *J. Power Sources* 196, 886–893. doi: 10.1016/j.jpowsour.2010.09.031

95. Kulkarni, A., Ciacchi, F. T., Giddey, S., Munnings, C., Badwal, S. P. S., Kimpton, J. A., et al. (2012). Mixed ionic electronic conducting perovskite anode for direct carbon fuel cells. *Int. J. Hydrogen Energy* 37, 19092–19102. doi: 10.1016/j.ijhydene.2012.09.141

96. Kulkarni, A., and Giddey, S. (2013). Materials issues and recent developments in molten carbonate fuel cells. *J. Solid State Electrochem.* 16, 3123–3146. doi: 10.1007/s10008-012-1771-y

97. Kumar, B., Llorente, M., Froehlich, J., Dang, T., Sathrum, A., and Kubiak, C. P. (2012). Photochemical and photoelectrochemical reduction of CO_2. *Annu. Rev. Phys. Chem.* 63, 541–69. doi: 10.1146/annurev-physchem-032511-143759

98. Kurzweil, P. (2010). Gaston Planté and his invention of the lead–acid battery—the genesis of the first practical rechargeable battery. *J. Power Sources* 195, 4424–4434. doi: 10.1016/j.jpowsour.2009.12.126

99. Laguna-Bercero, M. A. (2012). Recent advances in high temperature electrolysis using solid oxide fuel cells: a review. *J. Power Sources* 203, 4–16. doi: 10.1016/j.jpowsour.2011.12.019

100. Lam, L. T., Haigh, N. P., Phyland, C. G., and Urban, A. J. (2004). Failure mode of valve-regulated lead-acid batteries under high-rate partial-state-of-charge operation. *J. Power*

Sources 133, 126–134. doi: 10.1016/j.jpowsour.2003.11.048

101. Lam, L. T., and Louey, R. (2006). Development of ultra-battery for hybrid-electric vehicle applications. *J. Power Sources*158, 1140–1148. doi: 10.1016/j.jpowsour.2006.03.022

102. Lee, A. C., Mitchell, R. E., and Gur, T. M. (2011). Feasibility of hydrogen production in a steam-carbon electrochemical cell.*Solid State Ionics* 192, 607–610. doi: 10.1016/j.ssi.2010.05.034

103. Lee, J., Kwon, Y., Machunda, R. L., and Lee, H. J. (2009). Electrocatalytic recycling of CO_2 and small organic molecules. *Chem. Asian J.* 4, 1516–1523. doi: 10.1002/asia.200900055

104. Leung, P. K., Ponce de Leon, C., Low, C. T. J., Shah, A. A., and Walsh, F. C. (2011b). Characterization of a zinc–cerium flow battery. *J. Power Sources* 196, 5174–5185. doi: 10.1016/j.jpowsour.2011.01.095

105. Leung, P. K., Ponce de Leon, C., and Walsh, F. C. (2011a). An undivided zinc–cerium redox flow battery operating at room temperature (295 K). *Electrochem. Commun.* 13, 770–773. doi: 10.1016/j.elecom.2011.04.011

106. Lex, P., and Jonshagen, B. (1999). The zinc/bromine battery system for utility and remote area applications. *Power Eng. J,* 13, 142–148. doi: 10.1049/pe:19990307

107. Li, W. (2010). "Electrocatalytic reduction of CO_2 to small organic molecule fuels on metal catalysts," in *Advances in CO_2Conversion and Utilization*, ed Y. Hu (Washington, DC: ACS Symposium Series, American Chemical Society), 55–76.

108. Google Scholar

109. Licht, S., Wang, B., Ghosh, S., Ayub, H., Jiang, D., and Ganley, J. (2010). A new solar carbon capture process: solar thermal electrochemical photo (STEP) carbon capture. *J. Phys. Chem. Lett.* 1, 2363–2368. doi: 10.1021/jz100829s

110. Lim, H., Lackner, A., and Knechtli, J. (1977). Zinc-bromine secondary battery. *J. Electrochem. Soc.* 124, 1154. doi: 10.1149/1.2133517

111. Liu, Q., Shinkle, A., Li, Y., Monroe, C., Thompson, L., and Sleightholme, A. (2010). Non-aqueous chromium acetylacetonate electrolyte for redox flow batteries. *Electrochem. Commun.* 12, 1634–1637. doi: 10.1016/j. elecom.2010.09.013

112. Liu, Q., Sleightholme, A., Shinkle, A., Li, Y., and Thompson, L. (2009). Non-aqueous vanadium acetylacetonate electrolyte for redox flow batteries. *Electrochem. Commun.* 11, 2312–2315. doi: 10.1016/j.elecom.2009.10.006

113. Lodhi, M. A. K., and Daloglu, A. (2001). Design and material variation for an improved power output of AMTEC cells. *J. Power Sources* 93, 32–40. doi: 10.1016/S0378-7753(00)00538-3

114. Logan, B. E., and Rabaey, K. (2012). Conversion of wastes into bioelectricity and chemicals by using microbial electrochemical technologies. *Science* 337, 686–690. doi: 10.1126/science.1217412

115. Manthiram, Y. F., and Su, Y.-S. (2013). Challenges and prospects of lithium–sulfur batteries. *Acc. Chem. Res.* 46, 1125–1134. doi: 10.1021/ar300179v

116. Mo, Y., Ong, S. P., and Ceder, G. (2011). First-principles study of the oxygen evolution reaction of lithium-sir battery. *Phys. Rev. B* 84, 1–9. doi: 10.1103/PhysRevB.84.205446

117. Naoi, K., Naoi, W., Aoyagi, S., Miyamoto, J., and Kamino, T. (2013). New generation "Nanohybrid Supercapacitor." *Acc. Chem. Res.* 46, 1075–1083. doi: 10.1021/ar200308h

118. Narasimhaiah, G., and Janardhanan, V. M. (2013). Modeling CO_2 electrolysis in solid oxide electrolysis cell. *J Solid State Electrochem.* 17, 2361–2370. doi: 10.1007/s10008-013-2081-8

119. Nyugen, T., and Savinell, R. F. (2010). Flow batteries. The electrochemical society interface. *Fall* 2010, 54–56.

120. Ogura, K. (2013). Electrochemical reduction of carbon dioxide to ethylene: mechanistic approach. *J. CO_2 Util.* 1, 43–49. doi: 10.1016/j.jcou.2013.03.003

121. Paulenova, A., Creager, S., Navratil, J., and Wei, Y. (2002). Redox potentials and kinetics of the Ce^{3+}/Ce^{4+} redox reaction and solubility of cerium sulfates in sulfuric acid solutions. *J. Power Sources* 109, 431–438. doi: 10.1016/S0378-7753(02)00109-X

122. Phair, J. W., and Badwal, S. P. S. (2006a). Materials for separation membranes in hydrogen and oxygen production and future power generation. *Sci. Technol. Adv. Mater.* 7, 792–805. doi: 10.1016/j.stam.2006.11.005

123. Phair, J. W., and Badwal, S. P. S. (2006b). Review of proton conductors for hydrogen separation. *Ionics* 12, 103–115. doi: 10.1007/s11581-006-0016-4

124. Ponce de Leon, C., Frias-Ferrer, A., Gonzalez-Garcia, J., Szanto, D. A., and Walsh, F. C. (2006). Redox flow cells for energy conversion. *J. Power Sources* 160, 716–732. doi: 10.1016/j.jpowsour.2006.02.095

125. Qiao, J., Liu, Y., Hong, F., and Zhang, J. (2014). A review of catalysts for the electroreduction of carbon dioxide to produce low-carbon fuels. *Chem. Soc. Rev.* 43, 631–675. doi: 10.1039/c3cs60323g

126. Rabaey, K., Angenent, L., Schroder, U., and Keller, J. (2009). *Bioelectrochemical Systems: From Extracellular Electron Transfer to Biotechnological Application.* London: IWA publishing.

127. Reddy, T. B. (2011). *Linden's Handbook of Batteries. 4th Edn.* New York, NY: McGraw-Hill Education.

128. Remick, R. J., and Ang, P. G. P. (1984). *Electrically Rechargeable Anionically Active Reduction-Oxidation Electrical Storage Supply System.* U.S. Patent 4,485,154.

129. Ryan, M. A. (1999). "The alkali metal thermal-to-electric converter for Solar System exploration," in *Proceedings of Eighteenth International Conference on Thermoelectrics, 1999, Baltimore, USA* (New York, NY: IEEE), 630–638.

130. Salloum, K. S., and Posner, J. D. (2010). Counterflow membraneless microfluidic fuel cell. *J. Power Sources* 195,

6941–6944. doi: 10.1016/j.jpowsour.2010.03.096

131. Salloum, K. S., and Posner, J. D. (2011). Membraneless microfludic fuel cell stack. *J. Power Sources* 196, 1229–1234. doi: 10.1016/j.jpowsour.2010.08.069

132. Sbar, N. L., Podbelski, L., Yang, H. M., and Pease, B. (2012). Electrochromic dynamic windows for office buildings. *Int. J. Sustain. Built Environ.* 1, 125–139. doi: 10.1016/j.ijsbe.2012.09.001

133. Scrosati, B., Hassoun, J., and Sun, Y.-K. (2011). Lithium-ion batteries. A look into the future. *Energy Environ. Sci.* 4, 3287–3295. doi: 10.1039/c1ee01388b

134. Seehra, M. S., and Bollineni, S. (2009). Nanocarbon boosts energy-efficient hydrogen production in carbon-assisted water electrolysis. *Int. J. Hydrogen Energy* 34, 6078–6084. doi: 10.1016/j.ijhydene.2009.06.023

135. Shirasaki, K., Yamamura, T., Herai, T., and Shiokawa, Y. (2006b). Electrodeposition of uranium in dimethyl sulfoxide and its inhibition by acetylacetone as studied by EQCM. *J. Alloys Comp.* 418, 217–221. doi: 10.1016/j.jallcom.2005.10.059

136. Shirasaki, K., Yamamura, T., and Shiokawa, Y. (2006a). Electrolytic preparation, redox titration and stability of pentavalent state of uranyl tetraketonate in dimethyl sulfoxide. *J. Alloys Comp.* 408, 1296–1301. doi: 10.1016/j.jallcom.2005.04.124

137. Skyllas-Kazacos, M. (2003). Novel vanadium chloride/polyhalide redox flow battery. *J. Power Sources* 124, 299–302. doi: 10.1016/S0378-7753(03)00621-9

138. Skyllas-Kazacos, M. (2009). "Secondary batteries: redox flow battery—vanadium redox," in *Encyclopedia of Electrochemical Power Sources*, eds J. Garche, P. Moseley, Z. Ogumi, D. Rand, and B. Scrosati (New York, NY: Elsevier), 444–453.

139. Skyllas-Kazacos, M., Chakrabarti, M. H., Hajimolana, S. A., Mjalli, F. S., and Saleem, M. (2011). Progress in flow battery

research and development. *J. Electrochem. Soc.* 158, R55–R79. doi: 10.1149/1.3599565

140. Skyllas-Kazacos, M., and Grossmith, F. (1987). Efficient vanadium redox flow cell. *J. Electrochem. Soc.* 134, 2950–2953. doi: 10.1149/1.2100321

141. Skyllas-Kazacos, M., Kazacos, G., Poon, G., and Verseema, H. (2010). Recent advances with UNSW vanadium-based redox flow batteries. *Int. J. Energy Res.* 34, 182–189. doi: 10.1002/er.1658

142. Snook, G. A., Kao, P., and Best, A. S. (2011). Conducting-polymer-based supercapacitor devices and electrodes. *J. Power Sources* 196, 1–12. doi: 10.1016/j.jpowsour.2010.06.084

143. Song, M.-K., Cairns, E. J., and Zhang, Y. (2013). Lithium/Sulfur batteries with high specific energy: old challenges and new opportunities. *Nanoscale* 5, 2186–2204. doi: 10.1039/C2NR33044J

144. Stiegel, G. J., Bose, A. C., and Armstrong, P. A. (2014). *Development of Ion Transport Membrane (ITM) Oxygen Technology for Integration in IGCC and Other Advanced Power Generation Systems.* Available online at:http://www.netl.doe.gov/publications/factsheets/project/proj136.pdf (Accessed September 28, 2014).

145. Stoots, C. (2010). "Production of synthesis gas by high-temperature electrolysis of H_2O and CO_2 (coelectrolysis)," in*Sustainable Fuels from CO_2, H_2O, and Carbon-Free Energy* (New York, NY: Columbia University). Available online at:http://energy.columbia.edu/files/2012/11/Coelectrolysis-Rev-2.pdf (Accessed September 2, 2014).

146. Stoots, C. M. (2006). "High-temperature co-Electrolysis of H_2O and CO_2 for syngas production," in *Fuel Cell Seminar, Preprint, INL/CON-06-11719* (Honolulu, HI).

147. Sum, E., and Skyllas-Kazacos, M. (1985). A study of the V(II)/V(III) redox couple for redox flow cell applications. *J. Power Sources* 15, 179–190. doi: 10.1016/0378-7753(85)80071-9

148. Sundmacher, K., Rihko-Struckmann, L. K., and Galvita,

V. (2005). Solid electrolyte membrane reactors: status and trends.*Catal. Today* 104, 185–199. doi: 10.1016/j.cattod.2005.03.074

149. U.S. Energy Information Administration (2014). *Today in Energy (2014)*. U.S. Energy Information Administration. Available online at: http://www.eia.gov/todayinenergy/ (Accessed September 2, 2014).

150. Ursua, A., Gandia, L. M., and Sanchis, P. (2012). "Hydrogen production from water electrolysis: current status and future trends," in *Proceedings of the IEEE* (New York, NY: Electronics and Electrical Engineers, Inc.), 410–426.

151. Van Eerten-Jansen, M. C. A. A., Heijne, A. T., Buisman, C. J. N., and Hamelers, H. V. M. (2011). Microbial electrolysis cells for production of methane from CO_2: long-term performance and perspectives. *Int. J. Energy Res.* 36, 809–819. doi: 10.1002/er.1954

152. Wang, H., and Ren, Z. J. (2013). A comprehensive review of microbial electrochemical systems as a platform technology.*Biotechnol. Adv.* 31, 1796–1807. doi: 10.1016/j.biotechadv.2013.10.001

153. Weber, A. Z., Mench, M. M., Meyers, J. P., Ross, P. N., Gostick, J. T., and Liu, Q. (2011). Redox flow batteries: a review. *J. Appl. Electrochem.* 41, 1137–1164 doi: 10.1007/s10800-011-0348-2

154. Weber, N. (1974). A thermoelectric device based on beta-alumina solid electrolyte. *Energy Convers.* 14, 1–7. doi: 10.1016/0013-7480(74)90011-4

155. Wei, W., Cui, X., Chen, W., and Ivey, D. G. (2011). Manganese oxide-based materials as electrochemical supercapacitor electrodes. *Chem. Soc. Rev.* 40, 1697–1721. doi: 10.1039/c0cs00127a

156. Wei, Y., Yang, W., Caro, J., and Wang, H. (2013). Dense ceramic oxygen permeable membranes and catalytic membrane reactors. *Chem. Eng. J.* 220, 185–203. doi: 10.1016/j.cej.2013.01.048

157. Wen, Y., Cheng, J., Ning, S., and Yang, Y. (2009). Preliminary study on zinc–air battery using zinc regeneration electrolysis with propanol oxidation as a counter electrode reaction. *J. Power Sources* 188, 301–307. doi: 10.1016/j.jpowsour.2008.11.054

158. Wen, Y., Cheng, J., Xun, Y., Ma, P., and Yang, Y. S. (2008b). Bifunctional redox flow battery 2. V(III)/V(II)–l-cystine(O_2) system. *Electrochim. Acta* 53, 6018–6023. doi: 10.1016/j.electacta.2008.03.026

159. Wen, Y. H., Cheng, J., Ma, P. H., and Yang, Y. S. (2008a). Bifunctional redox flow battery-1 V(III)/V(II)–glyoxal(O_2) system. *Electrochim. Acta* 53, 3514–3522. doi: 10.1016/j.electacta.2007.11.073

160. WEO. (2012). *WEO-2012 Factsheets - World Energy, Outlook*. Available online at:http://www.worldenergyoutlook.org/media/weowebsite/2012/factsheets.pdf (Accessed September 2, 2014).

161. Whipple, D. T., and Kenis, P. J. A. (2010). Prospects of CO_2 utilization via direct heterogeneous electrochemical reduction. *J. Phys. Chem. Lett.* 1, 3451–3458. doi: 10.1021/jz1012627

162. Wilson, A., Ehrlich, B., Roberts, T., Yost, P., Melton, P., and Malin, N. (2012). *Cutting-Edge Windows that can be Tinted on Demand*. BuildingGreen.com. Energy Solutions. Available online at: http://www2.buildinggreen.com/blogs/cutting-edge-windows-can-be-tinted-demand (Accessed September 2, 2014).

163. Wu, S.-Y., Xiao, L., and Cao, Y.-D. (2009). A review on advances in alkali metal thermal to electric converters (AMTECs). *Int. J. Energy Res.* 33, 868–892. doi: 10.1002/er.1584

164. Xia, X., Hong-Tao, L., and Liu, Y. (2002). Studies of the feasibility of a Ce^{4+}/Ce^{3+}-V^{2+}/V^{3+} redox cell. *J. Electrochem. Soc.* 149, A426–A430. doi: 10.1149/1.1456534

165. Xu, Y., Wen, Y., Cheng, J., Cao, G., and Yang, Y. (2010). A

study of tiron in aqueous solutions for redox flow battery application. *Electrochim. Acta* 55, 715–720. doi: 10.1016/j. electacta.2009.09.031

166. Xue, F., Wang, Y., Wang, W., and Wang, X. (2008). Investigation on the electrode process of the Mn(II)/Mn(III) couple in redox flow battery. *Electrochim. Acta* 53, 6636–6642. doi: 10.1016/j.electacta.2008.04.040

167. Yamamura, T., Shiokawa, Y., Ikeda, Y., and Tomiyasu, H. (2002). Electrochemical investigation of tetravalent uranium β-diketones for active materials of all-uranium redox flow battery. *J. Nuclear Sci. Technol.* Suppl. 3, 445–448.

168. Yamamura, T., Shirasaki, K., Li, D., and Shiokawa, Y. (2006b). Electrochemical and spectroscopic investigations of uranium(III) with N,N,N′,N′-tetramethylmalonamide in DMF. *J. Alloys Comp.* 418, 139–144. doi: 10.1016/j. jallcom.2005.10.055

169. Yamamura, T., Shirasaki, K., Shiokawa, Y., Nakamura, Y., and Kim, Y. (2004). Characterization of tetraketone ligands for active materials of all-uranium redox flow battery. *J. Alloys Comp.* 374, 349–353. doi: 10.1016/j.jallcom.2003.11.117

170. Yamamura, T., Watanabe, N., and Shiokawa, Y. (2006a). Energy efficiency of neptunium redox battery in comparison with vanadium battery. *J. Alloys Comp.* 408, 1260–1266. doi: 10.1016/j.jallcom.2005.04.174

171. Yan, J. H., Li, W. S., and Zhan, Q. Y. (2004). Failure mechanism of valve-regulated lead–acid batteries under high-power cycling. *J. Power Sources* 133, 135–140. doi: 10.1016/j. jpowsour.2003.11.075

172. Yang, Z., Zhang, J., Kintner-Meyer, M. C. W., Lu, X., Choi, D., Lemmon, J. P., et al. (2011). Electrochemical energy storage for green grid. *Chem. Rev.* 111, 3577–3613. doi: 10.1021/ cr100290v

173. You, D., Zhang, H., and Chen, J. (2009). A simple model for the vanadium redox battery. *Electrochim. Acta* 54, 6827–6836. doi: 10.1016/j.electacta.2009.06.086

174. Zhan, Z., Kobsiriphat, J., Wilson, R., Pillai, M., Kim, I., and Barnett, S. A. (2009). Syngas production by coelectrolysis of CO_2/H_2O: the basis for a renewable energy cycle. *Energy Fuels* 23, 3089–3096. doi: 10.1021/ef900111f

175. Zhang, K., Sunarso, J., Shao, Z., Zhou, W., Sun, C., Wang, S., et al. (2011). Research progress and materials selection guidelines on mixed conducting perovskite-type ceramic membranes for oxygen production. *RSC Adv.* 1, 1661–1676. doi: 10.1039/c1ra00419k

176. Zhang, L., Cheng, J., Yang, Y., Wen, Y., Wang, X., and Cao, G. (2008). Study of zinc electrodes for single flow zinc/nickel battery application, *J. Power Sources* 179, 381–387. doi: 10.1016/j.jpowsour.2007.12.088

177. Zhang, L. L., and Zhao, X. S. (2009). Carbon-based materials as supercapacitors electrodes. *Chem. Soc. Rev.* 38, 2520–2531. doi: 10.1039/b813846j

178. Zhao, P., Zhang, H., Zhou, H., and Yi, B. (2005). Nickel foam and carbon felt applications for sodium polysulfide/bromine redox flow battery electrodes. *Electrochim. Acta* 51, 1091–1098. doi: 10.1016/j.electacta.2005.06.008

179. Zhou, H. T., Zhang, H. M., Zhao, P., and Yi, B. L. (2006). A comparative study of carbon felt and activated carbon based electrodes for sodium polysulfide/bromine redox flow battery. *Electrochim. Acta* 51, 6304–6312. doi: 10.1016/j.electacta.2006.03.106

The Future of Butyric Acid in Industry

Mohammed Dwidar[1], Jae-Yeon Park[2], Robert J. Mitchell[1], and Byoung-In Sang[3]

[1]School of Nano-Bioscience and Chemical Engineering, Ulsan National Institute of Science and Technology, Ulsan, Republic of Korea

[2]New Renewable Energy Lab, SK Innovation Global Technology, Seoul, Republic of Korea

[3]Department of Applied Chemical Engineering, Hanyang University, Seoul, Republic of Korea

ABSTRACT

In this paper, the different applications of butyric acid and its current and future production status are highlighted, with a particular emphasis on the biofuels industry. As such, this paper discusses different issues regarding butyric acid fermentations and provides suggestions for future improvements and their approaches.

BUTYRIC ACID IN BIOFUELS AND OTHER INDUSTRIAL APPLICATIONS

Butyric acid has many uses in different industries, and currently there is a great interest in using it as a precursor to biofuels. Due to increases in petroleum prices as well as a continuous reduction in petroleum availability and a growing need for clean energy sources, research has recently been directed towards alternative fuel sources. Biofuels in general offer many advantages including sustainability, a reduction of greenhouse gas emissions, and security of supply. The term biofuel generally refers to solid, liquid, or gaseous fuels that are predominantly produced from biomass and will be used as such throughout this paper. Furthermore, liquid biofuels can be broadly classified into (a) bioalcohols, (b) vegetable oils and biodiesels, and (c) biocrude and synthetic oils [1].

As it is one of the most promising biofuels for replacing gasoline in the future, a lot of attention these days is paid toward biobutanol. Its primary use is as an industrial solvent, but it also offers several advantages over ethanol as a transportation fuel. For instance, the three carbon-carbon bonds in butanol provide more energy when burned than the two bonds present in two molecules of ethanol, that is, four carbons total for each fuel. In addition, butanol is less volatile than ethanol, can replace gasoline in internal combustion engines without any mechanical modifications, does not attract water like ethanol so it can be transported in existing pipelines, is not miscible with water, and is less sensitive to colder temperatures [2].

Despite these benefits, the fermentative bioproduction of butanol faces many problems as this alcohol is much more toxic to Clostridia than ethanol is to Zymomonas mobilis, which in turn results in lower concentrations in the fermentation broth, lower yields of butanol, and higher recovery costs. One feasible strategy to reduce the toxicity and improving the yield of butanol is to first ferment biomass into butyric acid and then convert this downstream into butanol [2]. In addition, butyric acid can be used to produce ethyl butyrate and butyl butyrate, both of which can be used as fuels [3].

In addition to its use as a biofuel, butyric acid has also many applications in pharmaceutical and chemical industries. Firstly, butyric acid is well known for its anticancer effects as it induces morphological and biochemical differentiation in a variety of cells leading to concomitant suppression of neoplastic properties [4–7]. Consequently, these studies present on various prodrugs that are derivatives of butyric acid were tried for their potential use in treatment of cancers and hemoglobinopathies, including leukemia and sickle cell anemia (SCA), and also to protect hair follicles of radio- and chemotherapy-induced alopecia. Butyrate also helps protect colonic mucosa from oxidative stress and inhibits its inflammation while promoting satiety [6]. In chemical industries, the main application of butyric acid is in the manufacture of cellulose acetate butyrate plastics [8]. By introducing the butyryl group into cellulose acetate polymers, the resulting polymer exhibits better performance in terms of its solubility in organic solvents due to enhanced hydrophobicity, better flexibility, and light and cold resistance [8]. A recent study also showed the possibility of mixing cellulose acetate butyrate (CAB) with another polymer, poly (3-hydroxybutyrate) (PHB), which can also be prepared from butyric acid [9], to decrease the production cost of PHB and improve its characteristics [10]. Furthermore, although butyric acid itself has an unpleasant odor, butyric acid esters such as methyl, ethyl, and amyl butyrate are used as fragrant and flavoring agents in beverages, foods and cosmetic industries [11, 12].

CURRENT STATUS OF BUTYRIC ACID PRODUCTION AND FUTURE NEEDS

Butyric acid is produced at the industrial scale via mainly a chemical synthesis. This involves the oxidation of butyraldehyde, which is obtained from propylene derived from crude oil by oxosynthesis [13]. The chemical synthesis of butyric acid is preferred mainly because of its lower production cost and the availability of the starting materials. Another method for butyric acid preparation is its extraction from butter. This is possible since its concentration in butter ranges from 2% to 4% but the process involved is difficult and expensive and, thus, cannot compete with the chemical alternative [14]. A third way is through fermentation. Although this method is currently more expensive compared to the chemical synthesis, it has garnered more attention due to both a growing consumer desire for organic and natural products, as opposed to chemically synthesized ingredients, and a continuous increase in the prices of crude oil, which is needed for the chemical synthesis as noted earlier.

STRAINS AVAILABLE FOR THE BIOPRODUCTION OF BUTYRIC ACID

Under anaerobic conditions, butyric acid is a common metabolite produced by bacteria strains from various genera. However, for industrial use, Clostridial strains are preferred owing to their higher productivities and the final concentrations obtained. The most important Clostridium strains studied for industrial scale productions of butyric acid are C. butyricum [15–17], C. tyrobutyricum [18–24], and C. thermobutyricum [25] (Table 1). Currently, the most promising microorganism used for the bioproduction of butyric acid is C. tyrobutyricum. This strain is capable of producing butyric acid with high selectivity and can tolerate high concentrations of this compound. However, it can only ferment very few carbohydrates, including glucose, xylose, fructose, and lactate, while its ability to utilize mannitol or glycerol is doubtful [26, 27]. On the other hand, C. butyricum can ferment many carbon sources including hexoses, pentoses, glycerol, lignocellulose, molasses, potato starch, and cheese-whey permeate [17]. However, compared to C. tyrobutyricum, the butyrate yields have always been much lower (Table 1). For C. thermobutyricum, the range of fermentable sugars is somewhat between these other two strains as it consumes glucose, fructose, maltose, xylose, ribose, and cellobiose, and some oligomeric and polymeric sugars but not sucrose or starch [28].

Table 1: Summary of the most promising butyrate producing clostridial strains

Strain	Sugar	Final conc. (g/L)	Fermentation mode	Reference
C. butyricum ZJUCB	Glucose	12.25	Batch	[17]
		16.74	Fed-batch	
C. butyricum S21	Glucose	7.3	Batch	
	Sucrose	10	Extractive batch	[16]
	Sucrose	20	Pertractive fed-batch	
C. butyricum S21	Lactose	18.6	Batch	[13]
C. thermobu-tyricum ATCC 49875	Glucose	10.04	Batch	[25]
		19.38	Continuous	
C. beijerinckii	Lactose	12	Batch	[29]
C. popule-ti ATCC 35295	Glucose	6.3	Batch	[30]
C. tyrobutyricum JM1	Glucose	13.76	Batch	[19]

C. tyrobutyricum CIP 1–776	Glucose	45	Batch	[18]
	Glucose	62.8	Fed-batch	
	Glucose	28.6	Fed-batch	[21]
C. tyrobutyricum ATCC 25755	Glucose	24.88	Fed-batch (immobilized cells)	[20]
	Glucose	43.4	Fed-batch (immobilized cells)	[22]
	Glucose	53	Fed-batch (immobilized cells)	[23]
C. tyrobutyricum CNRZ 596	Glucose	44	Batch	
	Glucose	16.8	Continuous	[24]
	Glucose	33	Continuous (cell recycle)	
C. tyrobutyricum ZJU 8235	Jerusalem artichoke hydrolysate	27.5 g/L	Batch	[31]
		60.4 g/L	Fed-batch (immobilized cells)	

BUTYRIC ACID BIOSYNTHESIS IN CLOSTRIDIUM AND FACTORS AFFECTING ITS PRODUCTION

For the fermentative process to proceed, glucose must first be converted to pyruvate via the Embden-Meyerhof-Parnas (EMP) pathway, which produces two moles of ATP and NADH. Subsequently, pyruvate is fermented to produce several products, depending on the strain. Lactate dehydrogenase catalyzes the conversion of pyruvate to lactate while pyruvate-ferredoxin oxidorecuctase catalyzes its conversion to acetyl-CoA with the release of CO_2 and the generation of reduced ferredoxin (fdH_2). For acetate production, phosphotransacetylase (PTA) and acetate kinase (AK) are the key enzymes while, for butyrate, phosphotransbutyrylase (PTB) and butyrate kinase (BK) play similar roles (Figure 1). During acetate production, 4 moles of ATP are formed, while during butyrate production only 3 moles of ATP are formed, which helps to explain why, at high growth rates, cells shift more towards acetate rather than butyrate production [32, 33]. At the end of the exponential growth, the organisms slow down acetate production and take up excreted acetate, converting it into butyrate. This recycling may be an attempt of the organism to detoxify the medium by reducing the total hydrogen ion concentration, that is, acid concentration, as one butyric acid is produced from two acetic acids. Consequently, the metabolism is shifted from the more energy conserving acetate formation (1) to a lower acid content with butyrate formation (2) [9]:

$$\text{Glucose 2 Acetate} + 4H_2 + 2CO_2 + 4ATP \qquad (1)$$

$$\text{Glucose Butyrate} + 2H_2 + 2CO_2 + 3ATP. \qquad (2)$$

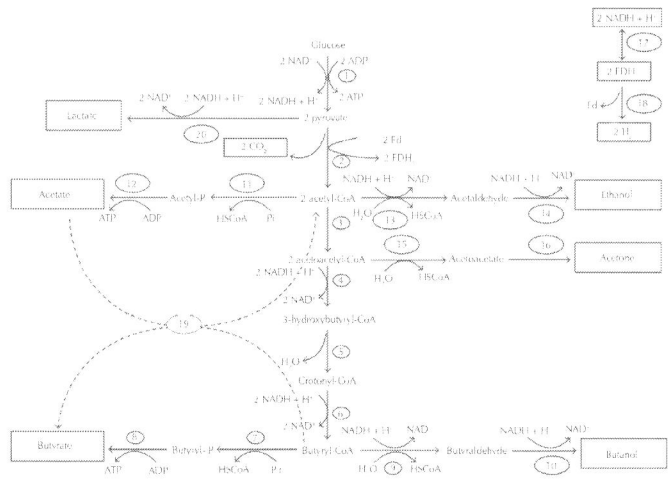

Figure 1: Metabolic pathway of butyrate production in Clostridia [19, 34, 35], 1: EMP pathway, 2: pyruvate-ferredoxin oxidoreductase, 3: acetyl CoA-acetyl transferase (thiolase), 4: -hydroxy butyryl CoA dehydrogenase, 5: crotonase, 6: butyryl CoA dehydrogenase, 7: phosphotransbutyrylase, 8: butyrate kinase, 9: butyraldehyde deydrogenase, 10: butanol dehydrogenase, 11: phosphotransacetylase, 12: acetate kinase, 13: acetaldehyde dehydrogenase, 14: ethanol dehydrogenase, 15: CoA transferase, 16: acetoacetate decarboxylase, 17: ferredoxin-NAD(P)+ reductase, 18: hydrogenase, 19: butyryl CoA-acetate transferase (proposed enzyme), 20: lactate dehydrogenase.

For those strains capable of producing solvents (butanol and acetone), the fermentation usually passes through two steps—the acidogenesis phase in which butyric and acetic acids are both produced in the medium and then the solventogenesis phase in which the organism converts these acids into acetone, ethanol, and butanol [14]. This second stage is initiated as the medium becomes more acidic and the cells enter the stationary phase [14].

Factors Affecting Butyrate Production and the Acetate/Butyrate Ratio

There are many factors that can affect the production characteristics. In general, butyrate production is higher in fed-batch, glucose-limited, and slow-growing cultures than in classic batch cultures [32, 36]. Likewise, the addition of acetate can significantly change the final butyrate concentrations. For example, it was found that the addition of acetate to cultures of C. thermobutyricum leads to higher final optical densities, a higher consumption of glucose, and faster growth rates [37]. Furthermore, in another study done recently [33], the addition of acetate (10 g/L) to a continuous culture of C. thermobutyricum led to 20% increase in the average concentration of butyric acid in the culture (from 10.23 g/L to 12.27 g/L). A similar phenomenon also was observed for C. tyrobutyricum CNRZ 59 [38]. Also for C. beijerinckii NCIMB 8052, it was found that cultures grown in medium containing added acetate exhibited higher CoA transferase, acetate kinase, and butyrate kinase activities than cultures grown in its absence [39].

The nutritional value of the medium may also have a great effect on the productivity and final butyrate concentration. Usually, butyric acid production is enhanced in rich medium. For C. tyrobutyricum, it gave more butyric acid in RCM media (which has more yeast extract and tryptone) compared to the Clostridial growth media (CGM) [23]. For C. thermobutyricum, it was found that its growth and glucose utilization is dependent on the yeast extract concentration [37]. Furthermore, peptone or tryptone was not adequate substitute for yeast extract and their addition to a

0.5% glucose-containing medium (to substitute for yeast extract) caused a strong increase in the frequency of sporulation. Yeast extract is also necessary for the growth of the butyrate-producing thermophile C. thermopalmarium [40]. These effects of yeast extract are common among other Clostridia [41] but are not as surprising as yeast extract is a complex nutritional source consisting of amino nitrogen, vitamins and other unknown growth factors used by many microbial organisms [42].

Trace elements were found also to affect butyrate production, especially iron and phosphate. For C. butyricumgrown on glycerol as a carbon source, it was found that phosphate limitation caused an increase in butyrate yield and a decrease in acetate yield, leading to a diminished acetate/butyrate ratio while iron limitation caused the reverse [43]. Interestingly, however, both resulted in significant improvement in the 1, 3-propandiol production and a decrease in the H_2/CO_2 ratio. The authors assumed that the hydrogenase enzyme (the enzyme responsible for releasing hydrogen from reduced ferredoxin instead of storing it as NAD (P) H_2) of C. butyricum contains iron and that it is greatly affected by iron limitations. They claimed that iron and phosphate limitation results in greater availability of reducing equivalents ($NADH_2$) and this in turn favors the production of NADH-dependent products, that is, mainly 1,3 propanediol (which has higher capacity for NADH consumption) and to a lesser extent butyrate and ethanol. However, with regard to the butyrate pathway, an iron limitation probably results in a reduced butyrate kinase activity and, thus, inhibition of butyrate production [44].

A similar effect to phosphate limitation can be also achieved by the addition of a suitable electron mediator, such as methyl viologen, which can act as an artificial electron-carrier system like ferredoxin, accepting electrons coming from pyruvate-ferredoxin oxidoreductase. As the electrons bound to methyl viologen are less available for the hydrogenase than those coming from ferredoxin, a greater proportion of electrons are channeled to the ferredoxin-NAD(P) reductase to produce $NAD(P)H_2$ and thus the acetate/butyrate ratio will decrease [43, 45]. The activity of the hydrogenase can be also decreased through an increase in the hydrogen partial pressure [36, 43, and 46] or through carbon monoxide flushing (a reversible inhibitor of hydrogenase) [47]. In contrast to these studies, when Jo et al. [48] overexpressed the [Fe Fe]-hydrogenase in Clostridium tyrobutyricum JM1, their purpose being to enhance hydrogen production, they found that significant changes in the main metabolic pathways occurred, which led to an improved butyric acid yield while also suppressing lactic acid production [48]. Therefore, it seems that hydrogenases are key enzymes controlling the butyric, acetic, and lactic acids pathways and that their activities should be carefully tuned or monitored to achieve an optimum production.

Effect of Carbon Source and Fermentation Products

In general, butyric acid fermentations are more prolific in terms of their yields, productivity, and final butyrate concentrations when performed with limited glucose concentrations as compared to those done with excess glucose, as the latter often leads to osmotic dehydration of the cells [49]. However, end product inhibition is the main problem that faces butyric acid fermentations, and different approaches are being tried to improve the tolerance of Clostridium strains to the produced butyric acid. Butyric acid can pass through the bacterial membrane in its unionized form and then dissociates inside the cell leading to disturbance in the pH gradient across the cell membrane and thus more energy is consumed to restore and maintain a functional pH gradient across the membrane, which in turn will limit the energy available for biomass growth [38, 50].

It was observed that butyric acid produced by the culture is more toxic than that externally added, and the same seems to apply for acetic acid. This could be explained by the fact that acid concentration, within acid-producing cells, is higher than that when the acids are added externally [51]. Butyrate was found to cause more inhibition compared to acetate for C. tyrobutyricum and also for C. acetobutylicum, C. populeti, and C. thermocellum [32].

In general, the inhibitory effects of the byproducts (e.g., acetate, butanol, ethanol, and acetone) appear in a concentration range that is above the concentrations usually reached during butyric acid fermentations. For solventogenic strains, which convert the produced butyrate into butanol, butanol is the toxic end product and negatively impacts the culture through its fluidizing effect on the lipids of the cell membrane, leading to disruption of the membrane function [52].

APPROACHES FOR IMPROVING BUTYRIC ACID BIOPRODUCTION

Fermentation Approaches

Almost all known fermentation modes, ranging from batch and fed-batch to continuous fermentations with and without cell recycling, were evaluated during butyrate production by different research groups. For fed-batch fermentation, the rate of substrate supply is a key factor for optimizing the production. In 1989, Fayolle et al. reported that substrate feeding controlled by the rate of gas production was preferable to a constant rate feeding [18].

Compared to batch and fed-batch modes, continuous cultures can give higher productivities but lower product concentration. To enhance the productivity even more, cell recycling systems can be applied with a continuous process design, allowing operation at a high cell concentration and at a limited growth rate, and, consequently, a high butyrate production rate and selectivity can be achieved [24, 32, and 38]. Another way to maintain higher cell density and thus higher butyrate productivity is through cell immobilization [14, 53], but it should be kept in mind that immobilized cell bioreactors tend to lose their productivity over time due to the accumulation of old inactive cells. This drawback can be observed in various kinds of cell immobilization techniques [9]. However, fibrous bed bioreactors (FBBs) seem to be less affected by this phenomenon as compared to other systems and can maintain their productivity over a longer period of time if the operative conditions are carefully controlled [54]. This is attributed to their large void volume and high permeability, which give them regenerative ability and high mass transfer capabilities [20]. It should not come as a surprise, therefore, that FBBs have been successfully used by different research groups to produce butyric acid [20, 33, 55, and 56].

In Situ (Online) Product Removal

Online product removal through different techniques including dialysis, distillation, adsorption, and extraction was tried for isolation of carboxylic acids and other volatile products, and many of them can be applied for butyric acid fermentations [57]. In situ removal of the product can improve the fermentation process and enhances the productivity by decreasing the concentration of this product in the culturing medium and therefore reducing its toxic effect on the cells. Besides, this greatly reduces the need for the tedious and time consuming separation steps that follow the fermentation process.

For electrodialysis, the biomass should be separated first from the fermentation broth through micro- or ultrafilteration, and then the product of interest can be subsequently separated through permeate electrodialysis. This method was applied successfully for the separation of propionic, lactic, and acetic acids [58–60]. Pervaporation processes are also frequently used for isolation of volatile products, and they were successfully used for the in situ removal of solvents in ABE fermentations using C. acetobutylicum as a fermenting strain [61]. Also adsorption is fairly common for the isolation of fermentation products such as butanol using various materials as adsorbents [62, 63]. The use of extraction and pertraction methods was mentioned by many research groups for the isolation of products like organic acids and other fermentation products; however, several factors have to be considered before the butyrate fermentation is coupled with extraction or pertraction. Both physical and reactive extractions have been described and evaluated for their potential application in connection with fermentation. Reactive extraction in general has higher efficiency compared to the physical one because the organic phase also contains a reactant or carrier. It means that the acid is extracted into an organic phase by physical transport and complexation with the carrier. Several chemicals can be applied as the organic phase but the toxicity of the organic solvent that can be estimated from its Log P (partition coefficient) value must be considered first together with its extractive ability [64]. One way to decrease this toxicity is through immobilizing the cells and thus protecting them from being in close contact with the organic solvent [22]. Further protection can be achieved also through adding vegetable oils such as castor, soy bean, and olive oils to the immobilization matrix to trap the diffusing solvent molecules in the fermentation medium [65]. For example, some authors reported the use of a mixture of Al_2O_3 and sunflower oil to decrease the toxicity of decanol in extractive fermentation for ethanol production using S. cerevisiae cells immobilized in calcium alginate gel, and they claimed the possibility of applying similar strategies for other valuable chemicals also [66]. Similarly, another research group reported the possibility of decreasing the toxicity of alamine 336/oleyl-alcohol extraction system on Lactobacillus delbrueckii through immobilization and adding soybean oil to the -carrageenan matrix [67]. The effectiveness of the extraction process and the distribution coefficient depend greatly on pH as the organic solvents can extract only undissociated acids. This problem can be partially solved by reactive extraction or by pertraction in which the product is extracted from the fermentation broth and simultaneously stripped from the organic phase into the stripping solution. The organic phase is simultaneously regenerated in this process.

Extraction and pertraction were used effectively for acetate, propionate, and lactate fermentations and also for ABE fermentation. For butyrate fermentation, Zigova et al. in 1996 [68] tested different solvents including tertiary amines, alcohols, alkanes, and vegetable oils for their potential use as extractants, and they found that solvents with tertiary amines were the best in terms of their butyric acid extractive abilities. The toxicity also of various solvents for C. butyricum was evaluated by the same research group [69]. In 1999, they showed the possibility of integrating extraction and pertraction methods with fermentative production of butyric acid byC. butyricum. They used Hostarex (20% w:w) in oleylalcohol as an extractant, and the concentration of butyric acid produced was increased from 7.3 g/L in the control to 10.0 g/L in extractive fermentation and to 20.0 g/L in pertractive fermentation with concurrent increases in the yield also [16]. In 2003, Wu and Yang reported also on the use of pertractive fermentation for butyric acid production from glucose, using immobilized cells ofC. tyrobutyricum in a fibrous bed bioreactor. In their study, they used 10% (v/v) alamine 336 in oleyl alcohol as the extractant with simultaneous regeneration through continuous stripping with NaOH [22].

Recently, another research group reported on the use of a tertiary alkyl phosphine compound, tri-n-octylphosphine oxide (TOPO) (which is more environmentally friendly than amines) dissolved in seven different solvents for extraction of proionic and butyric acids from aqueous solutions [70]. It was observed that the use of TOPO dissolved in these diluents increased the distribution coefficients of propionic and butyric acids between the organic and aqueous phases up to 51.22 and 12 times, respectively. The results also showed that the highest distribution coefficient for both acids is obtained by TOPO dissolved in methyl tert-butyl ether (MTBE). It should be noted, however, that the extraction ability of different organic mixtures is highly affected by temperature [71].

Aside from the use of extraction in online butyric acid removal, it can still be used also off line for downstream purification of butyric acid from the fermentation broth. Wu et al. [3] reported on the possibility of selective extraction of butyric acid from the fermentation broth through salting out using inorganic salts as calcium chloride [3]. This "salting out" effect was very efficient to separate butyric acid from the simulated butyrate fermentation broth (which consisted of butyric acid and acetic acid in a concentration ratio of 4 : 1) so that the final ratio of butyric acid/acetic acid in the upper phase was 9.87 : 1.

Some authors also suggested a novel approach for selective separation of dilute products from simulatedClostridium fermentation broth through the application of cyclodextrins. The ability of cyclodextrins (CDs) to form crystalline insoluble complexes with specific organic compounds was used to separate the ABE fermentation products together with acetic and butyric acids separately. Cyclodextrins were thus shown to offer a new exciting possibility for downstream processing of those products [72].

As the production of butyric acid and carboxylic acids in general is usually done near the neutral pH and in the form of salts, which requires the addition of mineral acids after finishing the fermentaion for the separation of the free acid, some authors proposed an alternative approach for carboxylic acids production. This approach is through two-stage fermentation process in which the fermentation medium containing a lactate salt is first fermented with the appropriate strain to give the salt of the required acid and then a fermentable carbohydrate is added into the medium to make the second fermentation medium which can be fermented by a lactobacillus strain under conditions suitable for converting the carbohydrate to a lactate salt while converting the salt of the selected acid into the free acid, then separating this free acid and recycling the medium again [73].

Approaches for Strain Improvement

Conventional butyric acid fermentation process is not yet economically competitive as the final concentration of butyric acid in the fermentation broth is still much lower than the required values with low yield as well as low productivity. Besides, acetic acid is produced as a byproduct in the process that causes further reduction in butyrate yield and increases product recovery and purification costs. To improve the economics of the fermentation process, it is desirable to increase butyrate while reducing acetate production [9]. To increase butyrate production, genetic methods were applied to C. tyrobutyricum ATCC 25755 strain aiming to delete the acetate-producing pathway [21]. The mutants of C. tyrobutyricum with either pta (phosphotransacetylase) or ak (acetate kinase) genes knocked out were constructed. Compared with the wild type, butyric acid production by these mutants was improved with higher final product concentration and yield. Also, these mutants have better tolerance to butyric acid inhibition. The ack-deleted mutant also has improved hydrogen production. However, acetic acid production in these mutants was not significantly affected in the fermentations even though the mutants had a much lower AK activity and the PTA-AK pathway should have been impaired. Besides PTA and AK, the authors suggested the presence of other enzymes in C. tyrobutyricum, which can also produce acetate, for example, CoA transferase, which can catalyze the formation of acetate from acetyl-CoA [21].

Sillers et al. [74] also reported on the metabolic engineering of the nonsporulating, nonsolvent-producing C. acetobutylicum M5 strain (which has lost the pSOL1 megaplasmid containing aad and the acetone formation genes) [74]. In that study, they tried to combine thl (thiolase) overexpression with aad (alcohol/aldehyde dehydrogenase) overexpression aiming at enhancing butanol formation. While this approach succeeded in limiting the formation of acetate and ethanol, the butanol titers were not improved. However, they found that overexpressing thl led to significant increase in butyric acid production.

Recently, Jiang et al. [56] reported on the adaption of C. tyrobutyricum in immobilized cultures in an FBB [56]. This adaptation resulted in significantly reduced inhibition effects of butyric acid on specific growth rate and also on cellular activities of phosphotransbutyrylase (PTB) and ATPase, together with elevated intracellular pH. Using this adapted strain, they could reach a final butyric acid concentration of about 86.9 g/L. This may be further improved in the future if the molecular mechanisms leading to growth inhibition by butyric acid can be determined.

SUMMARY AND FUTURE RESEARCH

Nowadays, there is a great need for butyric acid produced by microbial fermentation in various industries. The main future challenge for this will depend upon increasing butyrate tolerance of Clostridia strains and decreasing the cost of culture media materials. Lignocellulosic materials have the potential to serve as a cost-effective source of raw material for the production of liquid fuels, including bioethanol and biobutanol, and various organic chemicals. The main obstacle that faces the use of lignocellulosic feedstocks for fuel production is the high cost of hydrolyzing cellulose into simple monomeric sugars.

One suggested approach for solving this problem is through coculturing a strain capable of degrading these carbohydrate polymers such as cellulose, hemicelluloses, and starch with a Closteridium strain capable of producing the biofuel or the carboxylic acid of interest [9]. For example, it may be possible to coculture C. thermocellum and C. thermobutyricum for the production of butyric acid using cellulose as a carbon source. C. thermocellum is capable of producing cellulase and hemicellulase and is capable of utilizing the hexoses, but not the pentose sugars, generated from cellulose and hemicellulose. Although the main secondary metabolite from C. thermocellum is ethanol, not butyric acid, it suffers from a low growth rate [41], allowing C. thermobutyricum, which ferments pentose and hexose sugars more quickly than C. thermocellum, to utilize the sugars liberated and generate butyric acid. The recombinant technology to develop a cellulolytic organism is also possible, especially for C. thermobutyricum as its culture temperature is similar to the temperature at which cellulase is most active [9]. On the other hand, Chang et al. [75] reported that the combination of aerobic Bacillus and anaerobic Clostridium may play the key role in the future of biofuel production from biomass since strains of Bacillus have high growth rates in general and they can secrete many extracellular sacchrification enzymes in the medium such as amylase, pectinase, protease, cellulase, and hemicellulase. Tran et al. [76] also have shown recently the potential of coculturing B. subtilis with C. botulinum for butanol production from starch [76]. Besides those ordinary carbon sources, other groups suggested also the possible use of substrates as alcohols for carboxylic acid production through biotransformation using strains asAcetobacter aceti (for butyric acid from butanol) [77] and Gluconobacter oxydans (for propionic acid from n-propanol) [78].

Butyric acid is a promising chemical that may hold the potential for tomorrow energy needs as it can be converted to butanol through biological transformation. Even with the final goal being a higher butanol productivity, strains producing butyric acid may hold the key as they are capable of producing 3~5 times more butyrate than the current maximum seen for butanol, for example, about 20 g/L. Consequently, these strains can be used to generate a surplus of butyric acid, which can be converted downstream as needed.

ACKNOWLEDGMENTS

This paper was supported in part by the National Research Foundation of Korea (NRF) (Grant no. NRF-2011-616-D00042), the Ministry of Education, Science and Technology (Grant no. NRF-2009-C1AAA001-2009-0093479), and the research fund of Hanyang University (HY-201100000000233-N). The authors are grateful for their support.

REFERENCES

1.   A. Demirbas, "Biofuels securing the planet's future energy needs," Energy Conversion and Management, vol. 50, no. 9, pp. 2239–2249, 2009.

2.   J. Ebert, "Biobutanol: the next big biofuel," Biomass Magazine, 2008.

3.   D. Wu, H. Chen, L. Jiang, J. Cai, Z. Xu, and P. Cen, "Efficient separation of butyric acid by an aqueous two-phase system with calcium chloride," Chinese Journal of Chemical Engineering, vol. 18, no. 4, pp. 533–537, 2010.

4.   Z. X. Chen and T. R. Breitman, "Tributyrin: a prodrug of butyric acid for potential clinical application in differentiation therapy," Cancer Research, vol. 54, no. 13, pp. 3494–3499, 1994.

5.   A. Rephaeli, R. Zhuk, and A. Nudelman, "Prodrugs of butyric acid from bench to bedside: synthetic design, mechanisms of action, and clinical applications," Drug Development Research, vol. 50, no. 3-4, pp. 379–391, 2000.

6.   H. M. Hamer, D. Jonkers, K. Venema, S. Vanhoutvin, F. J. Troost, and R. J. Brummer, "Review article: the role of butyrate on colonic function," Alimentary Pharmacology and Therapeutics, vol. 27, no. 2, pp. 104–119, 2008.

7.   J. Leavitt, J. C. Barrett, B. D. Crawford, and P. O. P. Ts'o, "Butyric acid suppression of the in vitro neoplastic state of Syrian hamster cells," Nature, vol. 271, no. 5642, pp. 262–265, 1978.

8.   Y. Cao, H. Q. Li, and J. Zhang, "Homogeneous synthesis and characterization of cellulose acetate butyrate (CAB) in 1-Allyl-3-methylimidazolium chloride (AmimCl) ionic liquid," Industrial & Engineering Chemistry Research, vol. 50, no. 13, pp. 7808–7814, 2011.

9.   C. Zhang, H. Yang, F. Yang, and Y. Ma, "Current progress on butyric acid production by fermentation,"Current Microbiology, vol. 59, no. 6, pp. 656–663, 2009.

10.   E. El-Shafee, G. R. Saad, and S. M. Fahmy, "Miscibility, crystallization and phase structure of poly(3-hydroxybutyrate)/cellulose acetate butyrate blends," European Polymer Journal, vol. 37, no. 10, pp. 2091–2104, 2001.

11.   D. W. Armstrong and H. Yamazaki, "Natural flavours production: a biotechnological approach," Trends in Biotechnology, vol. 4, no. 10, pp. 264–267, 1986.

12. C. Shu, J. Cai, L. Huang, X. Zhu, and Z. Xu, "Biocatalytic production of ethyl butyrate from butyric acid with immobilized Candida rugosa lipase on cotton cloth," Journal of Molecular Catalysis B, vol. 72, no. 3-4, pp. 139–144, 2011.

13. R. Cascone, "Biobutanol—a replacement for bioethanol?" Chemical Engineering Progress, vol. 104, no. 8, pp. S4–S9, 2008.

14. J. Zigova and E. Sturdlk, "Advances in biotechnological production of butyric acid," Journal of Industrial Microbiology and Biotechnology, vol. 24, no. 3, pp. 153–160, 2000.

15. D. Vandak, M. Tomaška, J. Zigova, and E. Sturdlk, "Effect of growth supplements and whey pretreatment on butyric acid production by Clostridium butyricum," World Journal of Microbiology & Biotechnology, vol. 11, no. 3, p. 363, 1995.

16. J. Zigova, E. Sturdlk, D. Vandak, and S. Schlosser, "Butyric acid production by Clostridium butyricum with integrated extraction and pertraction," Process Biochemistry, vol. 34, no. 8, pp. 835–843, 1999.

17. G. Q. He, Q. Kong, Q. H. Chen, and H. Ruan, "Batch and fed-batch production of butyric acid by Clostridium butyricum ZJUCB," Journal of Zhejiang University, vol. 6, no. 11, pp. 1076–1080, 2005.   ·

18. F. Fayolle, R. Marchal, and D. Ballerini, "Effect of controlled substrate feeding on butyric acid production by Clostridium tyrobutyricum," Journal of Industrial Microbiology, vol. 6, no. 3, pp. 179–183, 1990.

19. J. H. Jo, D. S. Lee, J. Kim, and J. M. Park, "Effect of initial glucose concentrations of carbon and energy balances in hydrogen-producing Clostridium tyrobutyricum JM1," Journal of Microbiology and Biotechnology, vol. 19, no. 3, pp. 291–298, 2009.

20. L. Jiang, J. Wang, S. Liang, X. Wang, P. Cen, and Z. Xu, "Production of butyric acid from glucose and xylose with immobilized cells of clostridium tyrobutyricum in a fibrous-

bed bioreactor," Applied Biochemistry and Biotechnology, vol. 160, no. 2, pp. 350–359, 2010.

21. X. Liu, Y. Zhu, and S. T. Yang, "Butyric acid and hydrogen production by Clostridium tyrobutyricum ATCC 25755 and mutants," Enzyme and Microbial Technology, vol. 38, no. 3-4, pp. 521–528, 2006.

22. Z. T. Wu and S. T. Yang, "Extractive fermentation for butyric acid production from glucose by Clostridium tyrobutyricum," Biotechnology and Bioengineering, vol. 82, no. 1, pp. 93–102, 2003.

23. R. J. Mitchell, J. S. Kim, B. S. Jeon, and B. I. Sang, "Continuous hydrogen and butyric acid fermentation by immobilized Clostridium tyrobutyricum ATCC 25755: effects of the glucose concentration and hydraulic retention time," Bioresource Technology, vol. 100, no. 21, pp. 5352–5355, 2009.

24. D. Michel-Savin, R. Marchal, and J. P. Vandecasteele, "Butyric fermentation: metabolic behaviour and production performance of Clostridium tyrobutyricum in a continuous culture with cell recycle,"Applied Microbiology and Biotechnology, vol. 34, no. 2, pp. 172–177, 1990.

25. F. Canganella and J. Wiegel, "Continuous cultivation of Clostridium thermobutyricum in a rotary fermentor system," Journal of Industrial Microbiology and Biotechnology, vol. 24, no. 1, pp. 7–13, 2000.

26. M. P. Bryant and L. A. Burkey, "The characteristics of lactate-fermenting sporeforming anaerobes from silage," Journal of Bacteriology, vol. 71, no. 1, pp. 43–46, 1956.

27. T. Gibson, "Clostridia in Silage," Journal of Applied Bacteriology, vol. 28, no. 1, pp. 56–62, 1965.

28. J. Wiegel, S. U. Kuk, and G. W. Kohring, "Clostridium thermobutyricum sp. nov., a moderate thermophile isolated from a cellulolytic culture, that produces butyrate as the major product,"International Journal of Systematic Bacteriology, vol. 39, no. 2, pp. 199–204, 1989.

29. S. Alam, D. Stevens, and R. Bajpai, "Production of butyric

acid by batch fermentation of cheese whey with Clostridium beijerinckii," Journal of Industrial Microbiology, vol. 2, no. 6, pp. 359–364, 1988.

30. G. B. Patel and B. J. Agnew, "Growth and butyric acid production by Clostridium pupuleti," Archives of Microbiology, vol. 150, no. 3, pp. 267–271, 1988.

31. J. Huang, J. Cai, J. Wang et al., "Efficient production of butyric acid from Jerusalem artichoke by immobilized Clostridium tyrobutyricum in a fibrous-bed bioreactor," Bioresource Technology, vol. 102, no. 4, pp. 3923–3926, 2011.

32. D. Michel-Savin, R. Marchal, and J. P. Vandecasteele, "Control of the selectivity of butyric acid production and improvement of fermentation performance with Clostridium tyrobutyricum," Applied Microbiology and Biotechnology, vol. 32, no. 4, pp. 387–392, 1990.

33. W. Li, H. J. Han, and C. H. Zhang, "Continuous butyric acid production by corn stalk immobilized Clostridium thermobutyricum cells," African Journal of Microbiology Research, vol. 5, no. 6, pp. 661–666, 2011.

34. S. Saint-Amans, L. Girbal, J. Andrade, K. Ahrens, and P. Soucaille, "Regulation of carbon and electron flow in Clostridium butyricum VPI 3266 grown on glucose-glycerol mixtures," Journal of Bacteriology, vol. 183, no. 5, pp. 1748–1754, 2001.

35. H. Huang, H. Liu, and Y. R. Gan, "Genetic modification of critical enzymes and involved genes in butanol biosynthesis from biomass," Biotechnology Advances, vol. 28, no. 5, pp. 651–657, 2010.

36. J. G. Van Andel, G. R. Zoutberg, P. M. Crabbendam, and A. M. Breure, "Glucose fermentation by Clostridium butyricum grown under a self-generated gas atmosphere in chemostat culture," Applied Microbiology and Biotechnology, vol. 23, no. 1, pp. 21–26, 1985.

37. F. Canganella, S. U. Kuk, H. Morgan, and J. Wiegel, "Clostridium

thermobutyricum: growth studies and stimulation of butyrate formation by acetate supplementation," Microbiological Research, vol. 157, no. 2, pp. 149–156, 2002.

38. D. Michel-Savin, R. Marchal, and J. P. Vandecasteele, "Butyrate production in continuous culture of Clostridium tyrobutyricum: effect of end-product inhibition," Applied Microbiology and Biotechnology, vol. 33, no. 2, pp. 127–131, 1990.

39. C. K. Chen and H. P. Blaschek, "Effect of acetate on molecular and physiological aspects of Clostridium beijerinckii NCIMB 8052 solvent production and strain degeneration," Applied and Environmental Microbiology, vol. 65, no. 2, pp. 499–505, 1999.

40. A. L. A. Soh, H. Ralambotiana, B. Ollivier, G. Prensier, E. Tine, and J. L. Garcia, "Clostridium thermopalmarium sp. nov., a moderately thermophilic butyrate-producing bacterium isolated from palm wine in Senegal," Systematic and Applied Microbiology, vol. 14, no. 2, pp. 135–139, 1991.

41. A. Geng, Y. He, C. Qian, X. Yan, and Z. Zhou, "Effect of key factors on hydrogen production from cellulose in a co-culture of Clostridium thermocellum and Clostridium thermopalmarium," Bioresource Technology, vol. 101, no. 11, pp. 4029–4033, 2010.

42. J. S. Smith, A. J. Hillier, and G. J. Lees, "The nature of the stimulation of the growth of Streptococcus lactis by yeast extract," Journal of Dairy Research, vol. 42, no. 1, pp. 123–138, 1975.

43. A. Reimann, H. Biebl, and W. D. Deckwer, "Influence of iron, phosphate and methyl viologen on glycerol fermentation of Clostridium butyricum," Applied Microbiology and Biotechnology, vol. 45, no. 1-2, pp. 47–50, 1996.

44. R. Twarog and R. S. Wolfe, "Enzymatic phosphorylation of butyrate," The Journal of Biological Chemistry, vol. 237, no. 8, pp. 2474–2477, 1962.

45. S. Peguin and P. Soucaille, "Modulation of carbon and

electron flow in Clostridium acetobutylicum by iron limitation and methyl viologen addition," Applied and Environmental Microbiology, vol. 61, no. 1, pp. 403–405, 1995.

46. R. J. Lamed, J. H. Lobos, and T. M. Su, "Effects of stirring and hydrogen on fermentation products of Clostridium thermocellum," Applied and Environmental Microbiology, vol. 54, no. 5, pp. 1216–1221, 1988.

47. R. Datta and J. G. Zeikus, "Modulation of acetone-butanol-ethanol fermentation by carbon monoxide and organic acids," Applied and Environmental Microbiology, vol. 49, no. 3, pp. 522–529, 1985.

48. J. H. Jo, C. O. Jeon, S. Y. Lee, D. S. Lee, and J. M. Park, "Molecular characterization and homologous overexpression of [FeFe]-hydrogenase in Clostridium tyrobutyricum JM1," International Journal of Hydrogen Energy, vol. 35, no. 3, pp. 1065–1073, 2010.

49. V. H. Edwards, "The influence of high substrate concentrations on microbial kinetics," Biotechnology and Bioengineering, vol. 12, no. 5, pp. 679–712, 1970.

50. P. J. Henderson, "Ion transport by energy-conserving biological membranes," Annual Review of Microbiology, vol. 25, pp. 393–428, 1971.

51. A. P. Zeng, A. Ross, H. Biebl, C. Tag, B. Gunzel, and W. D. Deckwer, "Multiple product inhibition and growth modeling of Clostridium butyricum and Klebsiella pneumoniae in glycerol fermentation,"Biotechnology and Bioengineering, vol. 44, no. 8, pp. 902–911, 1994.

52. F. Baut, M. Fick, M. L. Viriot, J. C. Andre, and J. M. Engasser, "Investigation of acetone butanol-ethanol fermentation by fluorescence," Applied Microbiology and Biotechnology, vol. 41, no. 5, pp. 551–555, 1994.

53. K. Hanaki, S. Hirunmasuwan, and T. Matsuo, "Selective use of microorganisms in anaerobic treatment processes by application of immobilization," Water Research, vol. 28, no.

4, pp. 993–996, 1994.

54. Y. L. Huang, Z. Wu, L. Zhang, C. Ming Cheung, and S. T. Yang, "Production of carboxylic acids from hydrolyzed corn meal by immobilized cell fermentation in a fibrous-bed bioreactor," Bioresource Technology, vol. 82, no. 1, pp. 51–59, 2002.

55. X. Liu and S. T. Yang, "Kinetics of butyric acid fermentation of glucose and xylose by Clostridium tyrobutyricum wild type and mutant," Process Biochemistry, vol. 41, no. 4, pp. 801–808, 2006.

56. L. Jiang, J. Wang, S. Liang et al., "Enhanced butyric acid tolerance and bioproduction by Clostridium tyrobutyricum immobilized in a fibrous bed bioreactor," Biotechnology and Bioengineering, vol. 108, no. 1, pp. 31–40, 2011.

57. A. Freeman, J. M. Woodley, and M. D. Lilly, "In situ product removal as a tool for bioprocessing,"Bio/Technology, vol. 11, no. 9, pp. 1007–1012, 1993.

58. P. Boyaval, J. Seta, and C. Gavach, "Concentrated propionic acid production by electrodialysis," Enzyme and Microbial Technology, vol. 15, no. 8, pp. 683–686, 1993.

59. V. Habova, K. Melzoch, M. Rychtera, and B. Sekavova, "Electrodialysis as a useful technique for lactic acid separation from a model solution and a fermentation broth," Desalination, vol. 162, no. 1–3, pp. 361–372, 2004.

60. S. T. Zhang, H. Matsuoka, and K. Toda, "Production and recovery of propionic and acetic acids in electrodialysis culture of Propionibacterium shermanii," Journal of Fermentation and Bioengineering, vol. 75, no. 4, pp. 276–282, 1993.

61. A. H. Mollah and D. C. Stuckey, "Maximizing the production of acetone-butanol in an alginate bead fluidized bed reactor using Clostridium acetobutylicum," Journal of Chemical Technology and Biotechnology, vol. 56, no. 1, pp. 83–89, 1993.

62. I. S. Maddox, "Use of silicalite for the adsorption of n-butanol from fermentation liquors," Biotechnology Letters, vol. 4, no. 11, pp. 759–760, 1982.

63. A. B. Thompson, S. J. Cope, T. D. Swift, et al., "Adsorption of n-butanol from dilute aqueous solution with grafted calixarenes," Langmuir, vol. 27, no. 19, pp. 11990–11998, 2011.

64. A. Inoue and K. Horikoshi, "Estimation of solvent-tolerance of bacteria by the solvent parameter log P,"Journal of Fermentation and Bioengineering, vol. 71, no. 3, pp. 194–196, 1991.

65. H. Tanaka, S. Harada, H. Kurosawa, and M. Yajima, "Immobilized Cell System with Protection against Toxic Solvents," Biotechnology and Bioengineering, vol. 30, no. 1, pp. 22–30, 1987.

66. H. Kapucu and u. Mehmetoglu, "Strategies for reducing solvent toxicity in extractive ethanol fermentation," Applied Biochemistry and Biotechnology A, vol. 75, no. 2-3, pp. 205–214, 1998.

67. V. M. Yabannavar and D. I. C. Wang, "Strategies for reducing solvent toxicity in extractive fermentations," Biotechnology and Bioengineering, vol. 37, no. 8, pp. 716–722, 1991.

68. J. Zigova, D. Vandak, S. Schlosser, and E. Sturdik, "Extraction equilibria of butyric acid with organic solvents," Separation Science and Technology, vol. 31, no. 19, pp. 2671–2684, 1996.

69. D. Vandak, J. Zigova, E. Sturdlk, and S. Schlosser, "Evaluation of solvent and pH for extractive fermentation of butyric acid," Process Biochemistry, vol. 32, no. 3, pp. 245–251, 1997.

70. M. Bilgin, C. Arisoy, and I. Kirballar, "Extraction equilibria of propionic and butyric acids with tri-n-octylphosphine oxide/diluent systems," Journal of Chemical and Engineering Data, vol. 54, no. 11, pp. 3008–3013, 2009.

71. A. Keshav, K. L. Wasewar, and S. Chand, "Extraction of acrylic, propionic and butyric acid using aliquat 336 in oleyl alcohol: equilibria and effect of temperature," Industrial and Engineering Chemistry Research, vol. 48, no. 2, pp. 888–893, 2009.

72. H. Shity and R. Bar, "New approach for selective separation of dilute products from simulated clostridial fermentation broths using cyclodextrins," Biotechnology and Bioengineering, vol. 39, no. 4, pp. 462–466, 1992.

73. P. J. Brumm and R. Datta, Production of Organic Acids by an Improved Fermentation Process, US, 1989.

74. R. Sillers, A. Chow, B. Tracy, and E. T. Papoutsakis, "Metabolic engineering of the non-sporulating, non-solventogenic Clostridium acetobutylicum strain M5 to produce butanol without acetone demonstrate the robustness of the acid-formation pathways and the importance of the electron balance,"Metabolic Engineering, vol. 10, no. 6, pp. 321–332, 2008.

75. J. J. Chang, C. H. Chou, C. Y. Ho, W. E. Chen, J. J. Lay, and C. C. Huang, "Syntrophic co-culture of aerobic Bacillus and anaerobic Clostridium for bio-fuels and bio-hydrogen production," International Journal of Hydrogen Energy, vol. 33, no. 19, pp. 5137–5146, 2008.

76. H. T. M. Tran, B. Cheirsilp, B. Hodgson, and K. Umsakul, "Potential use of Bacillus subtilis in a co-culture with Clostridium butylicum for acetone-butanol-ethanol production from cassava starch,"Biochemical Engineering Journal, vol. 48, no. 2, pp. 260–267, 2010.

77. D. Druaux, G. Mangeot, A. Endrizzi, and J. M. Belin, "Bacterial bioconversion of primary aliphatic and aromatic alcohols into acids: effects of molecular structure and physico-chemical conditions," Journal of Chemical Technology and Biotechnology, vol. 68, no. 2, pp. 214–218, 1997.

78. J. Svitel and E. Sturdlk, "n-Propanol conversion to propionic acid by Gluconobacter oxydans," Enzyme and Microbial Technology, vol. 17, no. 6, pp. 546–550, 1995.

Chapter 3

Knowledge and Perceptions of Energy Alternatives, Carbon and Spatial Footprints, and Future Energy Preferences within a University Community in Northeastern US

Joanna Burger[1] and Michael Gochfeld[2]

[1]Division of Life Sciences, Consortium for Risk Evaluation with Stakeholder Participation, Environmental and Occupational Health Sciences Institute, Rutgers University, Piscataway, USA

[2]Consortium for Risk Evaluation with Stakeholder Participation, Environmental and Occupational Health Sciences Institute,

Environmental and Occupational Medicine, Rutgers Medical School, Rutgers University, Piscataway, USA

ABSTRACT

Our overall research aim was to examine whether people distinguished between the spatial footprint and carbon footprint of different energy sources, and whether their overall "worry" about energy types was related to future developed of these types. We surveyed 451 people within a university community regarding knowledge about different energy sources with regard to renewability and spatial and carbon footprints and attitudes about which energy type(s) should be developed further. Findings were: 1) Gas, oil and coal were rated as the least renewable, and wind, solar and hydro as the most renewable; 2) Oil and coal were rated as having the largest carbon footprint, while wind, solar and tidal were rated the lowest; 3) There were smaller differences in ratings for spatial footprints, probably reflecting unfamiliarity with the concept, although oil and gas were rated the highest; 4) Energy sources viewed as renewable were favored for future development compared with non-renewable energy sources, and coal and oil were rated the lowest; 5) Worry-free sources such as solar were favored; and 6) There were some age-related differences, but they were small, and there were no gender-related differences. Overall, subjects knew more about carbon footprints than spatial footprints, generally correctly identified renewable and non-renewable sources, and wanted future energy development for energy sources which were less worried about (e.g. solar, wind). These perceptions require in-depth examination in a large sample from different areas of the country.

INTRODUCTION

Public knowledge about environmental issues can affect attitudes and beliefs about pollution, development, and environmental protection [1,2]. Recently, many environmental concerns have focused on energy, renewable energy options, and the environmental costs of different energy options [3-5]. Carbon footprints have received great attention, but ecological footprints have received less. The calculation of ecological footprint of fuel types is complicated and consists of three main components: area needed for energy production (including mining and processing), area needed to sequester emissions of greenhouse gases, and the area needed for safe deposition of nitrogen, sulphur and other waste products [6,7]. These usually translate into carbon equivalent emissions, using global warming potential recommended by the International Panel on Climate Change [8]. Calculations of carbon equivalent emissions quickly lead to discussions of sustainability, production capabilities, and alternative fuels [7, 9-11].

These discussions have involved the public, and there are assessments of how the public views energy sources and renewable energy [2, 12-15]. Many papers examine one type of energy or another, or report support for renewable energy in general [2, 13]. Nuclear power has received much attention because of controversy surrounding safety, environmental risk and public opposition [16, 17]. With nuclear, siting issues, population density, accidents and emergency routes are concerns [18- 22], as large as concerns about proximity to nuclear facilities [16,23-24]. While the carbon footprint of different energy sources has figured prominently in these discussions [3,5,8], spatial footprint has not. That is, perceptions of the actual size required for different energy sources have not been examined.

Our overall aim was to explore whether people understood the relative size of the spatial footprint (and carbon footprint) of different energy sources, how much they worried about different energy sources, and whether their worry was related to which energy sources they thought should be further developed. We define spatial footprint as the actual physical space needed to support a given energy type—how much land is required for a wind or solar farm, or how much land is required for a nuclear power plant or a hydroelectric plant? This paper also examines the hypothesis that there is a relationship between perceptions of possible harm (personal worry) and the energy sources favored for development. Six questions are addressed: 1) What are perceptions of the relative size of the spatial footprint of different energy sources; 2) What are perceptions of the carbon footprint of different energy sources; 3) Which energy sources are renewable; 4) What is their overall worry rating for each energy source; 5) Which energy sources would they like to see developed; and 6) Are there any age-related differences in these perceptions?

We surveyed 451 students and non-students in a university community in central New Jersey in 2011. The energy sources listed in the survey were natural gas, nuclear, coal, solar, wind, tidal, hydro, oil, and geothermal, although worry was not addressed for the last two. We test the null hypotheses that: 1) there are no significant differences in perceptions about spatial and carbon footprints among energy sources; 2) there are no significant differences in ratings for energy sources to be further developed; 3) there are no age-related differences in these perceptions; and 4) there is no relationship between overall worry and energy sources to be further developed. Any findings apply to the study population, sampled at one time, and are meant to serve as a basis for further study in other communities and countries. Our data thus reflect local, rather than global perceptions, and thus can be related to local development or lack thereof. Even when people support particular technologies, they often do not accept them within their own community [e.g. 23, 25], although Greenberg [2] reported that people living near nuclear facilities favored more development of nuclear than the general population. Greenberg also found age-related differences in that older respondents were more likely to support increasing reliance on coal, gas, oil and nuclear power than younger respondents. For this reason, we examined age in our study.

Information on the public's views deal with perceptions or worries about renewable and non-renewable energy, rather than on their knowledge base [2,16]. Dalton et al. [26] surveyed tourist attitudes about renewable energy use in a hotel, and found that about 50% favored renewable energy, such as wind, but wanted to see onshore rather than offshore development. There is often a gap between perceptions of preferred energy types, and siting acceptance [23,27]. Others have focused on economic valuation of land for sustainable development [28].

Social trust is critical in risk/benefit decisions about environmental safety and health [20,29], but so is knowledge. Based on 239 published studies, Beierle [30] found that involving stakeholders in decisions resulted in higherquality decisions, but only if the public had a sufficient information base about alternatives. Reversing public opposition requires both understanding of public views and knowledge about the issues, as well as appropriate steps to obtain public approval [24,30], although knowledge does not always change attitudes [15,23,25]. Understanding public perceptions and knowledge about energy sources is a first step in involving the public and other stakeholders in decision-making, leading to better environmental decisions [30,31]. While positive perceptions of energy types may not lead to acceptance of facilities at a local level [25,32], information on future energy type preferences and perceptions of worry can inform decision-makers. Other concerns, such as housing values, noise, and unsightliness, also influence personal decisions [13, 33,34].

The concept of ecological footprints is older than that of carbon footprints, and deals with a resource accounting tool that measures how much productive land and sea is appropriated for a given human use (e.g. the footprint) [35]. In this paper spatial footprint is used to denote the physical space needed to operate a given energy source.

METHODS

The overall protocol was to interview students (aged 18 - 22; N = 196) and others (over age 22; N = 255) living and working in a university and surrounding community (restaurants, bus stops) in central New Jersey, to examine knowledge and views of different energy options. Interviews took place from 1 April to 15 May 2011.

Subjects were selected by approaching the first person encountered, and then approaching the third or fourth person encountered thereafter. Although this approach is not completely random, there is no reason to assume biases. Interviewers identified themselves as from Rutgers University, gave a brief description of the study, and answered all questions following the interview. Refusal rate was less than 5%, and people refused because they were late for class or other appointments, had small children, or were rushing to board a bus. The interviews required about 20 minutes; many were longer due to subject's questions or comments about energy and politics which were allowed after the survey was completed. The protocol was approved as exempt by the Rutgers Institutional Review Board.

The questionnaire contained 4 parts concerning: 1) the relative size of the carbon footprint and the spatial footprint [on a scale of 1-5]; 2) whether each energy source was renewable or not; 3) how much they worried about different energy sources and favored further development of each energy source for the United States; and 4) demographics. Carbon footprint was defined as the relative amount of carbon emissions per kilowatt hour (kwh) of electric output, and spatial footprint was defined as how much land was required per kwh. Demographics included gender and age. A pilot survey of 10 students indicated that their ratings were not significantly different on two different days. There were no clear gender differences using Kruskal-Wallis non-parametric Analysis of Variance ($P > 0.10$ for comparisons), so gender is not discussed further.

We focused on nine energy sources: coal, oil, natural gas, solar, wind, tidal, nuclear, geothermal and hydroelectric. On a scale of 1 to 5, respondents were asked in a forced choice manner to rate from "low" to "high" the size of the carbon footprint and spatial footprint separately for each energy source. Respondents were also asked separately whether each energy source should be developed more in the US from "not at all" to "a lot" , and were asked "How worried are you about….," emphasizing individual rather than societal concern.

The "worry" question covered six of the energy types (oil, geothermal, and tidal were not on the worry question). Geothermal and tidal are not used in New Jersey and surrounding states, and oil accounts for only 1% of US electricity production [27]. The "worry" questions explored direct individual concerns including impacts on food, water, exposure of workers in the source facility, and exposure of wildlife from the facility. Exposure referred to radiation or radionuclides from nuclear facilities, mercury from coal-fired plants, carbon dioxide and sulphur emissions, and noise. Ratings were on a Likert Scale of 1 (no concern) to 5 (great concern). A composite worry index (mean score of the different worries) was computed from each energy source. On the pilot study, there was not a significant difference (Kruskal-Wallis tests, $P > 0.05$) on ratings for two different days.

Kruskal-Wallis non-parametric Analysis of Variance was used to compare estimates of renewability, spatial footprint, carbon footprint, worry and desirability across energy types, and also to analyze by age (up to age 22 versus 23 or older). A $P < 0.05$ was considered statisticcally significant, but readers should keep in mind the multiple comparisons inherent in this design.

RESULTS

Carbon and Spatial Footprint

Across the nine energy types there were significant differences in the estimates of the carbon and spatial footprint ratings (Figure 1, statistics given on figure). The differences were dramatic for carbon footprint, and subtle for spatial footprint. For carbon footprint, coal and oil were rated highest and hydro, wind, tidal and solar were rated the lowest, and results generally matched our own understanding. There were no age-related differences in the ratings. We concluded that most respondents had a basic understanding of carbon footprint.

For spatial footprint, there was little variation in the scores, hovering around 3 - 3.5, probably reflecting a lack of knowledge about spatial footprint—a concept generally ignored in media coverage. Indeed only 32% of responses gave a 1 or 5 compared with 54% for carbon footprint (2 = 7.12; P < 0.01). In general, subjects thought coal, nuclear, and oil had the largest spatial footprint, and geothermal, hydroelectric, and tidal had the smallest. The only age-related difference was for hydro, where older people thought hydro had a larger footprint than did younger people (Figure 1).

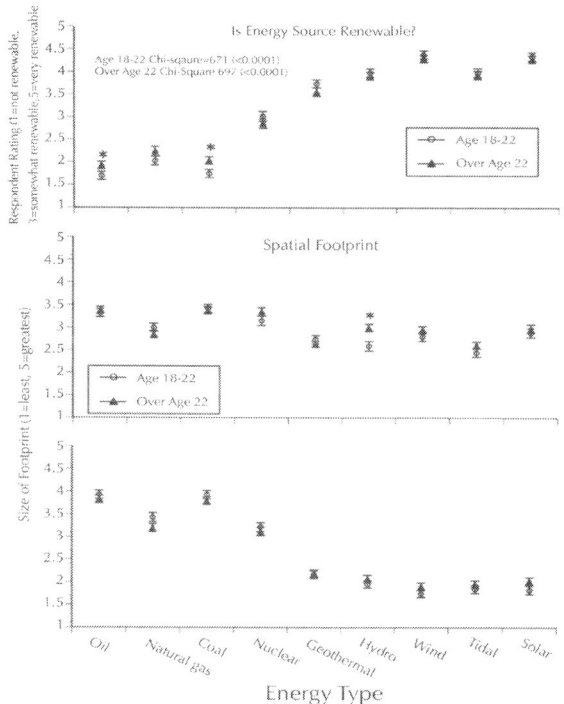

Figure 1: Ratings of people in a university community in central New Jersey about the relative size of the spatial and carbon footprints of different energy types. Shown also (bottom panel) are the ratings for whether an energy source is renewable or not. For the footprints, 1 = smallest, and 2 = largest. Shown are means ± standard error. Star equals significant age-related difference.

Renewability of Resources

There were significant differences in whether subjects rated energy sources as renewable or not (Figure 1, top panel). Generally, oil, gas and coal were rated as nonrenewable, with nuclear and geothermal in the middle. However, there was not a clear dichotomy; some respondents provided intermediate ratings. There were only two significant age-related differences. Older subjects rated oil and coal as more renewable than did students.

Energy Sources for Future Development

Subjects were asked to rate their views about which sources should be developed more in the United States. On Figure 2, we illustrate the percentage of subjects that rated each energy type a 1 ("do not develop" bottom panel) or a 5 (develop further, top panel). The energy source that had the lowest mean rating for future development was coal, followed by oil and nuclear, then by natural gas. The energy source with the highest rating was solar, followed by wind. Thus, people felt most strongly about developing solar, and not developing coal, than for the other energy sources. There were few age related differences. More young people were negative about oil and natural gas than were older people (Figure 2).

Overall Worry Rating

There were significant differences among energy type in the overall rating of worry (Figure 3, top panel). Nuclear had the highest worry index, followed by natural gas, and coal. People were less worried about wind, and solar (Figure 3). Generally, older people were more worried about more forms of energy (5 out of 6 categories, binomial 2-tailed $p = 0.125$) than were younger people, although the differences were significant only for nuclear. Although older subjects were more worried about nuclear, this was not reflected in the future development question.

The factors in the worry index included transportation risks, exposures from the plant or facility, exposure from food or water, exposure of workers, and exposure of wildlife. We also computed the mean worry score for each type of worry for all energy types combined.

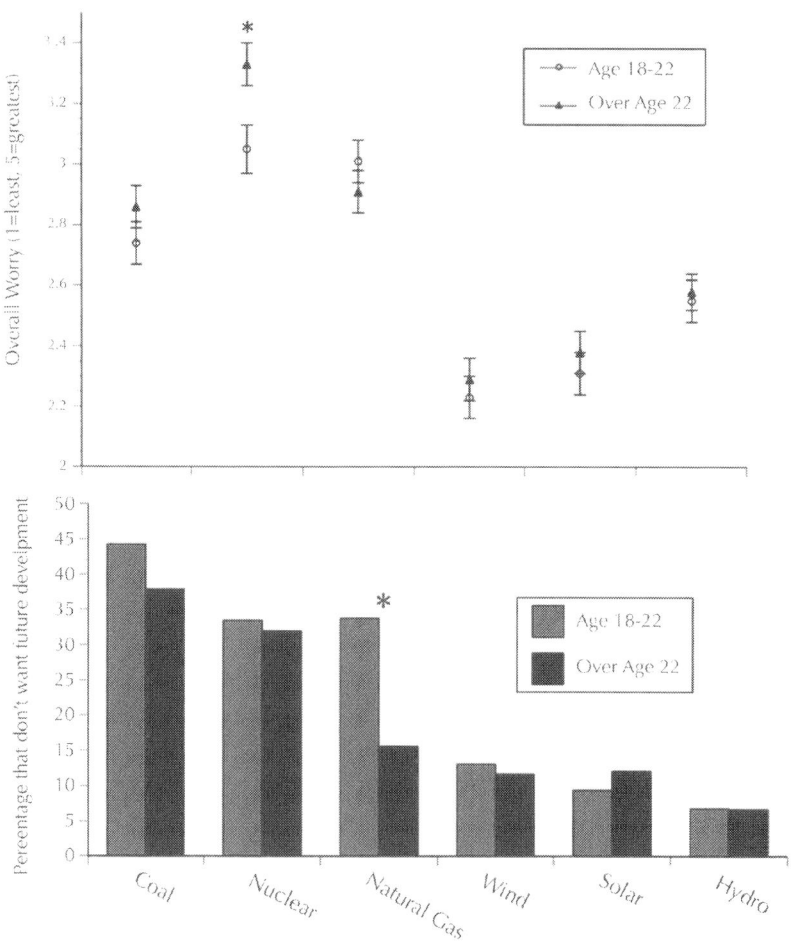

Figure 2: Percent of people wanting to see a particular energy type developed more in the future (rating of 5 of 5) and percent not wanting to see any future development (rating of 1 of 5). Star indicates significant age-related difference.

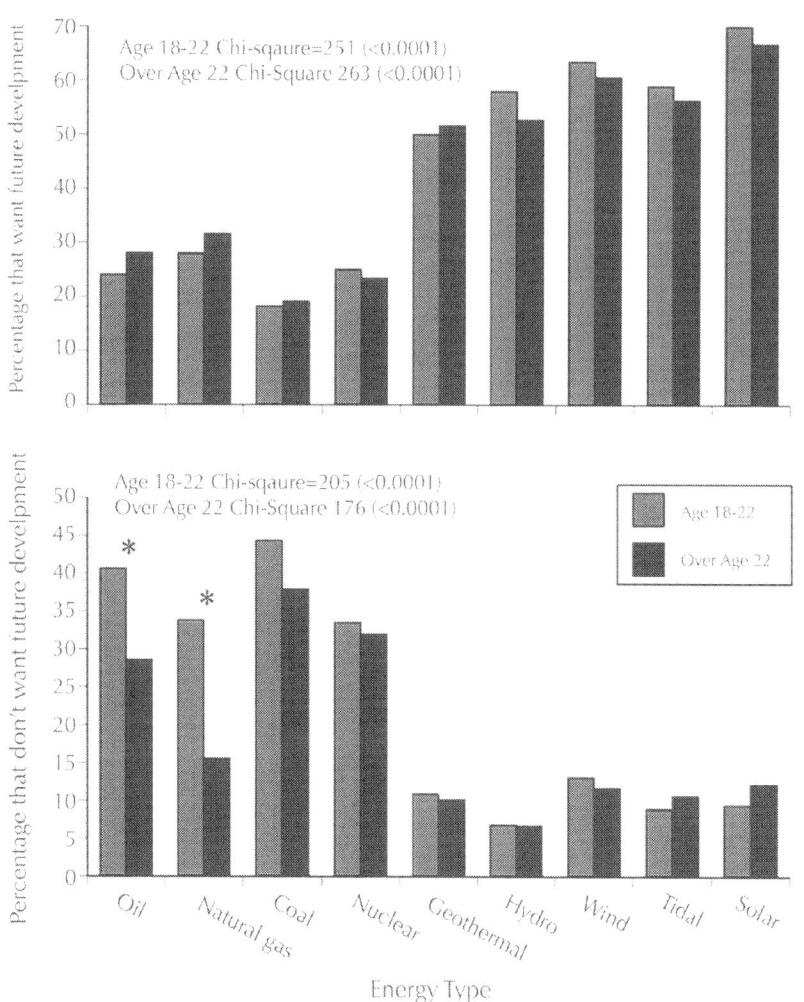

Figure 3: Overall worry score or index for different energy sources for college students (open circles) and people over 22 (black triangle). Shown are means ± standard error. Star equals significant age-related difference.

There were significant differences, with people being more worried about risks/exposures to food (mean of 3.12 ± 0.05) and wildlife (mean of 3.03 ± 0.05) than they were for workers (2.67 ± 0.05) and from the facility itself (2.49 ± 0.04). Other components scored even lower. Older respondents were significantly more worried about exposure from the plant itself, and from transportation, than were younger people (X^2 tests, $P < 0.05$).

The relationship between overall worry and energy types subjects did not want to develop, is also shown in Figure 3. Although generally related, people did not want to see coal developed further, even though they were less worried about coal than for either nuclear or natural gas. In contrast, people were more worried about hydro than their response on future development would suggest.

In summary, the null hypotheses were rejected with respect to all questions. There were significant differences as a function of energy type in rating of relative spatial and carbon footprint size, understanding of renewable energy, perceptions of which energy source to develop, and in overall worry about different energy forms. The relationship between overall worry and the percentage of people who do not want future development were generally related, except for coal (Figure 3). There were few age-related differences.

DISCUSSION

This study tested both knowledge and perceptions, and inevitably revealed some misconceptions regarding energy and climate. A 2011 report from the Yale University Center on Climate Policy reported that 90% of respondents identified development of clean energy and 70% rated global warming as medium, high, or very high priority for the country, even to the point that 65% supported a "carbon tax" [36]. A 2012 report for the World Energy Summit found that American concerns for energy security and the economic impacts of energy choices ran high, accompanying an interest in renewable and alternative energy sources [37]. The report reflected a strong interest in incentivizing renewable energy, based on both environmental and economic concerns [37]. Thus it is reasonable to predict that students and residents in a University town would have at least a basic understanding of these issues and their relationships. Further, one might expect younger people to have a greater understanding because they are still in school and exposed to some of these issues [see 2].

Carbon and Spatial Footprints

In this survey we asked specifically about "carbon footprint" and "spatial footprint" rather than ecological footprint, a more complex and controversial construct [8]. Understanding carbon footprints, reducing carbon emissions, and reversing global climate change is one of the foremost current ecological and media issues. Considerable attention has been given to examining global drivers, and to the need to reduce emissions from fossil-burning fuel (for electricity and transport) and industrial processes that have been accelerating rapidly [3,20]. The currency for these discussions is "carbon footprint', which relates to the amount of carbon released as carbon dioxide per unit (typically per kilowatt hour). Carbon is released by the burning of all types of fossil fuel and the carbon/kwh depends on the thermal density of the fuel, the efficiency of the combustion process, and air pollution control devices (although the latter only redirect the carbon from air to some other disposal process).

We expected respondents to have at least a basic grasp of the general issue of carbon associated with familiar energy types, since this has received extensive media attention. Respondent rankings correspond well to our own expectations, including the recognition that geothermal emits more carbon (in addition to sulfur) than other "renewable" sources [38]. Their ratings were generally correct despite the fact that neither hydrothermal nor hydro are used in New Jersey and thus the ratings do not reflect local experience.

Determining spatial footprints is difficult because of differences in the physical environment. This topic gets little media coverage, and not surprisingly respondents seemed unfamiliar with the concept, judging by their middle-of-the-road responses. A few examples of the complexities of spatial footprint will suffice: 1) slope, updrafts and local geography influence how many wind towers can be efficiently placed on a given amount of land, 2) the size and depth of the thermal field determines how much electricity can be generated from the field, and 3) weather patterns and latitude influence solar capacities, and how many solar cells are needed, facing in what directions, and how much energy is required to rotate them. For solar, the spatial footprint on a roof can be discounted compared to the usurpation of otherwise productive agricultural acreage or natural landscape. In the production of energy from biomass (not examined in this study), a water footprint must be considered since different plants (crops) require different amounts of water to produce a unit of energy [39], and the release of carbon from biomass burning varies by crop type [40].

Further, determination of both carbon and spatial footprints depend upon whether only direct footprints are considered, or indirect are as well. For example, the direct footprint of a hydropower generation plant includes the occupied area of the dam and plant, the build-up of land surrounding the facility, and the flooding of land behind a dam. Indirect effects include machinery production, building materials, what workers require to run the plant and the energy (either from hydro or fossil fuel) that is required for all machinery and materials to run to hydropower plant. Similarly, the direct footprint of a nuclear plant is the land area occupied by reactors, other buildings, storage pools and pads, as well as buffer areas. But the area involved in mining and processing, and ultimately the off-site disposal or reprocessing of spent fuel rods must be considered. Surface mining versus underground mining, and surface disposal (currently on site) versus repository storage of fuel, would provide different footprints.

Although there were differences among energy types in the spatial footprint responses, the responses reflect unfamiliarity rather than knowledge. That is, the ratings by subjects did not reflect current science. Few calculations have been made to compare with the perceptions reported in this paper. St glehner [9], however, provided some comparisons, and found that spatial footprints per energy produced decreased as follows: coal (highest spatial footprint, relative value of 20), oil (12), gas (10), biofuels, hydropower, solar, and wind (all less than 1). Geothermal, nuclear, and tidal were not examined. Huijbregts [41] provided another accounting of ecological footprint (in decreasing order) of biomass, hydro, wind and solar, and fossil and nuclear energy, but did not examine indirect footprints. Using energy chains for cars, Holden and H yer [7] came up with a ecological footprint ranking of biomass > oil > natural gas > hydro. The most inclusive ranking is from Sovacool [42]: coal > oil > natural gas >> nuclear > geothermal = biomass > solar = hydro = wind. The estimated release ranges from 1000 g CO_2/kWh for coal to about 10 g/kWh for solar, wind and hydro. There are several discrepancies depending on assumptions. For example, the full nuclear cycle includes substantial carbon emission in the front end, although negligible carbon is released during the reactor operations [42]. These discordant analyses illustrate the importance of scientists deciding on a uniform method of calculating spatial or ecological footprints.

In the present study of respondents from a university community in central New Jersey, the ranking of spatial footprint (in decreasing order) was: coal/oil/nuclear/ geothermal > hydro/wind/solar > tidal, but the differences were small with mean ratings for spatial footprint between 2.5 and 3.5, while they rated carbon footprints as varying from 1.8 to 4. Moreover, there were few age-related differences in knowledge about carbon and spatial footprints (refer to Figure 1). Older people thought the spatial footprint for hydro was larger than did younger people, but the differences were not great and may not be meaningful.

Renewability of Resources

Much of the public debate, public-policy decisions, and international agreements concern the dichotomy between renewable and non-renewable resources. Renewable resources are those that are naturally renewed, such as solar, wind, tidal, geothermal, and to varying extents biomass [43, 44]. Our definition of "renewable" is an energy source that is not depleted by use. In our view, there is not a perfect dichotomy, but instead there are intermediate stages. Geothermal, for example, in not completely renewable because it requires recharge to maintain the steam source [38].

Various polls have shown wide-spread support for the concept of renewable energy for environmental, economic, and security reasons [37]. Some people, however, have questioned whether the current high material living standards in developed nations; can be maintained using only renewable energy [45]. Several agencies and governments have addressed the development of energy plans [46] and systems that are 100% renewable [47,48], acknowledging that these would involve major societal changes in farming practices (if biofuels are key), use of land (if solar and wind), and possible offshore effects (if offshore wind), not to mention the direct environmental effects.

A reasonable public debate that leads to public policy decisions and the siting of energy facilities, however, requires an understanding of which sources are renewable, as well as the relative spatial footprint each requires. Clearly the most renewable energy source is solar, since the sun's energy striking the earth is relatively constant, taking into account latitude/season and atmospheric clarity, and wind energy which results from the sun's differential heating of the earth's surface. In the present study, subjects rated both solar and wind as the most renewable, although the average rating was less than 5, meaning that some people did not consider it completely renewable. At present, New Jersey has little solar or wind energy, although these are being encouraged by State government and the media.

Improved technology aims at increasing the efficiency of energy conversion for solar as well as for other forms of energy considered renewable (wind, tidal, geothermal, hydro), which requires energy-dependent generators to convert the renewable energy into electricity [49]. Geothermal, companies, for example, developed methods of powering the generators from geothermal energy rather than depending upon oil, but were slow in becoming independent [50]. Geothermal has the clear advantage of not influencing global warming [51].

We suggest that there are other distinctions that are rarely made when considering renewable resources—the degree of renewability and the predictability of the resource. For example, the sun will continue to shine, but wind is much less reliable, and geothermal is reliable but it can be overexploited. That is, if too much water is withdrawn from the geothermal field, the water table can drop (E. Gunniaugsson, Reykjavik Energy, Iceland, Pers. Comm.). Thus, there are complexities to the term "renewable" that require exploration. Further, methods of energy storage are critical for many forms of energy; the sun doesn't shine at night, and wind is not always strong enough to turn turbines [52].

Subjects in this survey correctly recognized solar and wind as renewable, and rated natural gas, coal and oil as non-renewable. Even so, however, everyone did not rate them a 1 (not renewable). Nuclear energy, usually considered non-renewable, but advantageous because of its low carbon emission, was rated as intermediate with a wide range of scores from 1 to 5. Thus, there seems to be less understanding among the respondents regarding the renewability status of nuclear. There were few age-related differences, and those that were significant were not great and may not be meaningful.

Greenberg and Truelove [53], in a survey of 3200 US residents, showed that there are multiple publics with respect to energy preferences and risk benefits. In our study, with a relatively homogeneous population within a university community, there was a wide difference in knowledge about the renewability of energy sources under discussion. It suggests public forums on energy resources and sustainability need to clearly define renewable, and identify the resources being discussed.

Worry, Knowledge and Energy Sources for Development

There is a very large literature on public preferences for, and worries about, different energy sources, with literally hundreds of opinion polls. Overall these polls show a clear preference for renewable sources of energy, and major reservations about coal and nuclear fuel [reviewed in 2]. Greenberg's national survey of 2701 US residents showed that over 90% wanted greater reliance on solar and wind, and over 70% wanted more reliance on hydroelectric sources. There is still concern, however, about the effect of wind on global climate [54].

In the present study, there was also a preference for wind and solar, followed by tidal, hydroelectric, and geothermal. Nuclear was more preferred for future development than natural gas, oil, and coal, which was surprising, given that the survey was conducted only weeks after the Fukushima nuclear event (March 2011), when the story was still receiving daily coverage in the media. In another series of questions, about 55% said that the Fukushima event and the Deep Water Horizon Gulf oil spill influenced their views about energy use (Burger, unpubl. data). Thus, it is likely that the Fukushima accident influenced the ratings, making it more surprising that nuclear was rated higher for future development than natural gas, oil, or coal.

The mean worry score for different energy sources was generally related to the percent of subjects who did not want that form of energy developed (Figure 3). However, this was only generally true. A higher percentage of subjects were opposed to further development of coal than their worry score would indicate. Generally subjects were not very worried about renewable energy forms (hydro, solar, wind), and few people opposed further development. Some people did, however, feel the renewables should not be further developed, and this bears further study.

Implications and Conclusions

Overall, subjects in this study had a reasonable understanding of the relative size of the carbon footprint, but less of an understanding of spatial footprints. The implications of this are that people may not be aware of the ecological consequences, in terms of physical space and the amount of ecosystems that would be disrupted, of different energy sources. It also suggests the importance of examining the relative physical impact of different energy sources on natural ecosystems. Understanding spatial footprint is particularly important for the state of New Jersey because it is a small state, with the highest population density in the US, where land is at a premium. We also suggest that permanence should be examined. That is, if a particular energy source is developed, can the ecosystem it replaces ever be restored once the energy source is developed? For example, could an ecosystem be restored if a wind farm or solar panels are placed there? This is an especially important question for New Jersey, where some farmland is being covered with solar panels.

Further, subjects correctly knew which energy sources were renewable and which were not, and they wanted to see more development of renewable resources, and less of non-renewable resources. Younger people wanted to see less future development of oil and natural gas than did older people, and a conclusion which agrees with the findings of Greenberg [2]. Thus, these results suggest that older people are less reluctant to move away from oil and natural gas, toward other forms of energy. However, when the data on the percentage of people who wish to see future development were examined (refer to Figure 2), there were no significant differences as a function of age. All age groups wanted to see future development of renewable energy sources, suggesting further support by people of efforts to develop renewable resources.

Worry can be used by managers to understand educational needs, and discrepancies between worry and their desire to forego further development of some energy sources. For example, people were less worried about coal than their preference for no more development would suggest. It seems people are not worried about it, but do not want to see further coal development. This also suggests that there is another reason for they wish not to see coal development that is not captured by their "worry" scores. The combination of preferences (or lack thereof) for future development, in conjunction with worry scores, may provide another way to examine personal perceptions of energy development.

Finally, this survey clearly indicated that people worry about the development of some energy sources (gas, oil, coal), and worry much less about others (wind and solar, followed by tidal, hydroelectric, and geothermal). The subjects interviewed generally wanted to see more future development of the energy sources that they were less worried about. The only energy source which did not fit this was nuclear (nuclear was more preferred for future development than natural gas, oil, and coal). Thus, overall, surveys can provide information on different aspects of future energy development, such as the public's rating of which sources to develop, their worry about different energy sources, and their knowledge (and worry) about carbon and ecological footprints.

ACKNOWLEDGEMENTS

This research was partly funded by the Consortium for Risk Evaluation with Stakeholder participation (DE-FC01- 06EW07053), and NIEHS Center Grant (P30ES005022), and EOHSI. The views expressed herein are solely those of the authors, and do not represent any of the funding agencies.

REFERENCES

1. M. Greenberg and K. Crossney, "The Changing Face of Public Concern about Pollution in the United States: A Case Study of New Jersey," Environment, Vol. 26, No. 4, 2006, pp. 255-268.

2. M. R. Greenberg, "Energy Sources, Public Policy, and Public Preferences: Analysis of US National and SiteSpecific Data," Energy Policy, Vol. 37, No. 8, 2009, pp. 3242-3249. doi:10.1016/j.enpol.2009.04.020

3. M. R. Raupach, G. Marland, P. Ciais, C. LeQuere, J. G. Canadell, G. Klepper and C. B. Field, "Global and Regional Drivers of Accelerating Cos Emissions," Proceedings of the

National Academy of Sciences, Vol. 104, No. 24, 2007, pp. 10288-10293.doi:10.1073/pnas.0700609104

4.   B. J. M deVries, D. P vanVuuren and M. M. Hoogwijk, "Renewable Energy Sources: Their Global Potential for the First Half of the 21st Century at the Global Level: An Integrated Approach," Energy Policy, Vol. 35, No. 4, 2007, pp. 2590-2610.doi:10.1016/j.enpol.2006.09.002

5.   EIA (Energy Information Administration), Energy-Related Carbon Dioxide Emissions, US Energy Information Administration, 2010. http://www.eia.gov/oiaf/ieo/emissions.html

6.   N. Chambers, C. Simmons and M. Wackernagel, "Sharing Nature's Interest: Ecological Footprints as an Indicator of Sustainability," Earthscan, London, 2000.

7.   E. Holden and K. G. H yer, "The Ecological Footprints of Fuels," Transportation Research, Part D, Vol. 10, No. 5, 2005, pp. 395-403.

8.   IPCC (International Panel on Climate Change), Climate Change 2007, Contribution of Working Group to the Fourth Assesment Report of the Intergovernmental Panel on Climage Change, Cambridge University Press, Cambridge, 2007.

9.   G. Stoglehner, "Ecological Footpring—A Tool for Assessing Sustainable Energy Supplies," Journal of Cleaner Production, Vol. 11, No. 3, 2003, pp. 267-277. doi:10.1016/S0959-6526(02)00046-X

10.   S. M. Benson and F. M. Orr Jr., "Sustainability and Energy Conversion" MRS Bulletin, Vol. 33, No. 4, 2008, pp. 297-302. doi:10.1557/mrs2008.257

11.   C. J. Bromley, M. Mongillo, G. Hiriart, B. Goldstein, R. Bertani, E. Huenges, A. Ragnarsson, J. Tester, H. Muraoka and V. Zui, V, "Contribution of Geothermal Energy to Climate Change Mitigation: The IPCC Renewable Energy Report," Proceedings of the World Geothermal Congress, Bali, 25-29 April 2010, pp. 1-5.

12. P. Upham, L. Whitmarsh, W. Poortinga, K. Purdam, A. Darnton, C. McLachlan and P. Devine-Wright, "Public Attitudes to Environmental Change: A Selective Review of Theory and Practice," Research Councils, Swindon, 2009. www.lwec.org.uk

13. J. Zoellner, P. Schweizer-Ries and C. Wemheurer, "Public Acceptance of Renewable Energies: Results from Case Studies in Germany," Energy Policy, Vol. 36, No. 11, 2008, pp. 4136-4141. doi:10.1016/j.enpol.2008.06.026

14. A. Spence, W. Poortinga, C. Butler and N. F. Pidgeon, "Perceptions of Climate Change and Willingness to Save Energy Related to Flood Experience," Nature, Vol. 1, 2011, pp. 46-49.

15. G. Ellis, J. Barry and C. Robinson, "Many Ways to Say 'No'-Different Ways to Say 'Yes'; Applying Q-Methodology to Understand Public Acceptance of Wind Farm Proposals," Journal of Environmental Planning and Management, Vol. 50, No. 4, 2007, pp. 517-551.doi:10.1080/09640560701402075

16. M. R. Greenberg, "How Much Do People Who Live near Major Nuclear Facilities Worry about Those Facilities: Analysis of National and Site-Specific Data," Journal of Environmental Planning and Management, Vol. 52, No. 7, 2009, pp. 919-937.doi:10.1080/09640560903181063

17. H. C. Hung and T. W. Wang, "Determinants and Mapping of Collective Perceptions of Technological Risk: The Case of the Second Nuclear Power Plant in Taiwan," Risk Analysis, Vol. 31, No. 4, 2011, pp. 668-682. doi:10.1111/j.1539-6924.2010.01539.x

18. R. R. Kasperson, O. Renn, P. Slovic, H. S. Brown, J. Emel, R. Goble, J. X. Kasperson and S. Ratick, "The Social Amplification of Risk: A Conceptual Framework," Risk Analysis, Vol. 8, No. 2, 1988, pp. 177-187. doi:10.1111/j.1539-6924.1988.tb01168.x

19. M. Wolsink, "Entanglement of Interests and Motives: Assumptions behind the NIMBY Theory on the Facility

Sitting," Urban Studies, Vol. 31, No. 6, 1994, pp. 851- 866. doi:10.1080/00420989420080711

20. G. O. Rogers, "Siting Potentially Hazardous Facilities: What Factors Impact Perceived and Acceptable Risk?" Landscape and Urban Planning, Vol. 39, No. 4, 1998, pp. 265-281. doi:10.1016/S0169-2046(97)00087-X

21. P. Slovic, "The Perceptions of Risk," In: J. Slovic, Ed., The Perception of Risk, Earthscan, London, 2000, pp. 221-230.

22. D. L. Feldman and R. A. Hanahan, "Public Perceptions of a Radioactively Contaminated Site: Concerns, Remediation Preferences, and Desired Involvement," Environmental Health Perspectives, Vol. 104, No. 12, 1996, pp. 1344-1352. doi:10.1289/ehp.961041344

23. M. R. Greenberg, "NIMBY, CLAMP, and the Location of New Nuclear-Related Facilities: US National and 11 Site Specific Surveys," Risk Analysis, Vol. 29, No. 9, 2009b, pp. 1242-1245. doi:10.1111/j.1539-6924.2009.01262.x

24. H. C. Jenkins-Smith, C. L. Silva, M. C. Nowlin and G. deLozier, "Reversing Nuclear Opposition: Evolving Public Acceptance of a Permanent Nuclear Waste Disposal Facility," Risk Analysis, Vol. 31, No. 4, 2011, pp. 629- 644. doi:10.1111/j.1539-6924.2010.01543.x

25. D. Bell, T. Gray and C. Haggett, "The 'Social' Gap in Wind Farm Sitting Decisions; Explanations and Policy Responses," Environmental Policy, 2005, pp. 49-64.

26. G. J. Dalton, D. A. Lockington and T. E. Baldock, "A Survey of Tourist Attitudes to Renewable Energy Supply in Australian Hotel Accommodation," Renewable Energy, Vol. 33, No. 10, 2008, pp. 2174-2185. doi:10.1016/j.renene.2007.12.016

27. United States Energy Information Administration (USEIA), US Carbon Dioxide Emissions from Energy Sources, EIA of DOE, 2010.

28. T. Soderqvist, H. Eggert, B. Olsson and A. Soutukorva, "Economic Valuation for Sustainable Development in the

Swedish Coastal Zone," Ambio, Vol. 34, No. 2, 2005, pp. 169-175.

29. M. Siegrist, G. Cvetkovih and C. Roth, "Salient Value Similarity, Social Trust, and Risk/Benefit Perception," Risk Analysis, Vol. 20, No. 3, 2000, pp. 353-362.doi:10.1111/0272-4332.203034

30. T. C Beierle, "The Quality of Stakeholder-Based Decisions," Risk Analysis, Vol. 22, No. 4, 2002, pp. 739-749. doi:10.1111/0272-4332.00065

31. T. Dietz and P. C. Stern, "Public Participation in Environmental Assessment and Decision-Making," National Academy Press, Washington DC, 2008.

32. P. Devine-Wright, "Local Aspects of UK Renewable Energy Development; Exploring Public Beliefs and Policy Implications," Local Environment, Vol. 10, No. 1, 2005, pp. 57-69.doi:10.1080/1354983042000309315

33. J. Blake, "Overcoming the 'Value-Action Gap' in Environmental Policy: Tensions between National Policy and Local Experience," Local Environ, Vol. 4, No. 3, 1999, pp. 257-278.doi:10.1080/13549839908725599

34. R. Kahn, "Siting Struggles; the Unique Challenge of Permitting Renewable Energy Power Plants," The Electric Journal, Vol. 13, No. 2, 2000, pp. 21-33. doi:10.1016/S1040-6190(00)00085-3

35. J. Kitzes, A. Peller, S. Goldfinger and M. Wachernagel, "Current Methods for Calculating National Ecological Footprint Accounts," Scientific Environmental Sustainability Society, Vol. 4, 2007, pp. 1-9.

36. A. Leiserowitz, E. Maibach, C. Roser-Renouf, N. Smith and J. D. Hmielowski, "Climate Change in the American Mind: Public Support for Climate & Energy Policies in November 2011," Yale University and George Mason University. New Haven, 2011.

37. Council on Foreign Relations, US Opinion on Energy Security, 2012.

38. J. L. Renner, "Geothermal Energy," In: T. M. Letcher, Ed., Future Energy: Improved, Sustainable and Clean Options for Our Planet, Elsevier, New York, 2008, pp. 211-224.

39. P. W. Gerbens-Leenes, A. Y. Hoekstra and T. H. van der Meer, "The Water Footprint from Biomass: A Quantitative Assessment and Consequences of an Increasing Share of Bio-Energy in Energy Supply," Ecological Economy, Vol. 68, No. 4, 2009, pp.1032-1060.doi:10.1016/j.ecolecon.2008.07.013

40. P. Champagne, "Biomass," In: T. M. Letcher, Ed., Future Energy: Improved, Sustainable and Clean Options for Our Planet, Elsevier, New York, 2008, pp. 151-170.

41. M. A. J. Huijbregt, S. Hellweg, R. Frischknecht, K. Hungerbuhler and A. J. Hendriks, "Ecological Footprint Accounting in the Life Cycle Assessment of Products," Ecology Economics, Vol. 64, No. 4, 2008, pp. 798-807. doi:10.1016/j.ecolecon.2007.04.017

42. B. Sovacool, "Valuing the Greenhouse Gas Emissions from Nuclear Power: A Critical Survey," Energy Policy, Vol. 36, No. 8, 2008, pp. 2940-2953.doi:10.1016/j.enpol.2008.04.017

43. D. M. Berman and J. T. O'Connor, "Who Owns the Sun? People, Politics and the Struggle for a Solar Economy," Chelsea Green Publishing Co., White River Junction, VT, 1996.

44. W. Shi, "Renewable Energy: Finding Solutions for a Greener Tomorrow," Reviews in Environmental Science and Biotechnology, Vol. 9, No. 1, 2010, pp. 33-37.doi:10.1007/s11157-010-9187-6

45. F. E. Trainer, "Can Renewable Energy Sources Sustain Affluent Society?" Energy Policy, Vol. 23, No. 12, 1995, pp. 1009-1026. doi:10.1016/0301-4215(95)00085-2

46. E. E. Thorhallsdottir, "Environment and Energy in Iceland: A Comparative Analysis of Values and Impacts," Environmental Impact Assessment Review, Vol. 27, No. 6, 2007, pp. 522-544. doi:10.1016/j.eiar.2006.12.004

47. H. Lund, "Renewable Energy Strategies for Sustainable Development," Energy, Vol. 32, No. 6, 2007, pp. 912-919. doi:10.1016/j.energy.2006.10.017

48. H. Lund and B. V. Mathiesen, "Energy System Analysis of 100% Renewable Energy Systems—the Case of Denmark in Years 2030 and 2050," Energy, Vol. 34, No. 5, 2009, pp. 524-531. doi:10.1016/j.energy.2008.04.003

49. V. Smil, "Energy Transitions: History, Requirements, Prospects," Praeger, California, 2010.

50. L. Rybach and M. Mongillo, "Geothermal Sustainability—A Review with Identified Research Needs," GRC Transaction, Vol. 30, 2006, pp. 1083-1090.

51. B. A. Goldstein, G. Hiriart, J. Tester, B. Bertani, R. Bromley, L. Guierrez-Negrin, C. J. Huenges, H. Ragnarsson, A. Mongillo, M. A. Muraoka and V. I. Zui, "Great Expectations for Geothermal Energy to 2100," Proceedings 36th Workshop of Geothermal Reservoir Engineering, Stanford, 31 January-2 February 2011.

52. D. Dicaire and F. H. Tezel, "Regeneration and Efficiency Characterization of Hybrid Adsorbent for Thermal Energy Storage of Excess and Solar Heat," Renewable Energy, Vol. 36, No. 3, 2011, pp. 986-992. doi:10.1016/j.renene.2010.08.031

53. M. Greenberg and H. B. Truelove, "Energy Choices and Risk Beliefs: It Is Just Global Warming and Fear of a Nuclear Power Plant Accident?" Risk Analysis, Vol. 31, No. 5, 2011, pp. 819-831. doi:10.1111/j.1539-6924.2010.01535.

54. D. W. Keith, J. F. DeCarolis, D. C. Denkenberger, D. H. Lenschow, S. L. Malyshev, S. Pacala and P. J. Rasch, "The Influence of Large-Scale Wind Power on Global Climate," Proceedings of the Natural Academy of Science of the United States of America, Vol. 101, No. 46, 2004, pp. 16115-16120. doi:10.1073/pnas.0406930101.

Chapter 4

Anaerobic Membrane Bioreactors (AnMBR) for Wastewater Treatment

Sheng Chang

School of Engineering, University of Guelph, Guelph, Canada

ABSTRACT

This paper focuses on the recent research in the development of anaerobic membrane bioreactors in wastewater treatment. Anaerobic wastewater treatment technology is gaining increasing attention due to its capacity to convert wastewater BODs to usable biogas with relatively low energy consumption. The anaerobic membrane bioreactor (AnMBR), which is a combination of the anaerobic biological wastewater treatment process and membrane filtration, represents a recent development in the high-rate anaerobic bioreactors. This paper reviews applications and performances of AnMBR and the membrane filtration behaviour in AnMBRs.

INTRODUCTION

Environmental sustainability is one of the most critical challenges that we are currently facing. To maintain a sustainable environment requires effective and advanced waste and wastewater management technologies which should not only remove the contaminants, but also be of high energy efficiency with the capacity to recover useful resources from waste and wastewater. One of the technologies to meet such requirements is the anaerobic digestion which can convert the waste BOD to usable biogas, reserve useful nitrogen and phosphorus for further recovery, and require minimum energy to operate. However, the efficiency of anaerobic digestion has been largely limited by the intrinsic slow growth rate of the anaerobic microorganisms, which results in a large reactor volume necessary for wastewater treatment by anaerobic digestion. One of the advanced solutions to improve the efficiency of anaerobic treatment is to integrate anaerobic wastewater treatment reactor with membrane filtration process to form an anaerobic membrane bioreactor system. In this system, the membrane filtration process can separate the treated wastewater from the anaerobic biomass and, at the same time, concentrate the biomass concentration in the anaerobic bioreactor to a desired level.

Although the concept of AnMBR was developed in 1980s [1], applications of the anaerobic membrane technology have been limited by concerns on the membrane fouling in the anaerobic environment, the energy consumption of the membrane processes, and the pre-matured large-scale wastewater treatment membrane filtration technology. However, with the success of MBR technology in recent years, the large-scale membrane filtration systems have become a proved technology with effective strategies for the membrane process design, operation, and maintenance developed for the biological wastewater treatment applications. These progresses and the potential of the AnMBR as an energy recovery technology have stimulated increased research interests in AnMBRs. This paper reviews the current status and recent development.

ANMBR SYSTEMS

The anaerobic membrane bioreactor is an integrated system of the anaerobic bioreactor and the low pressure ultrafiltration or microfiltration membrane filtration. Since MF/UF membranes can physically retain suspended solids, including suspended biomass and inert solids, the AnMBR can achieve complete separation of the solid retention time from the hydraulic retention time, independent of the wastewater characteristics, biological process conditions, and the sludge properties. As shown in Figure 1, the membrane filtration can be integrated with anaerobic bioreactors in three different forms: the internal submerged membrane filtration (A), the external submerged membrane filtration (B), and the external cross flow membrane filtration (C). The anaerobic bioreactor can be the complete mix [2], the up-flow anaerobic sludge blank (UASB) [3], the expanded granular sludge bed (EGSB) [4], the fluidized anaerobic bed reactors [5], and other type of anaerobic reactors. The complete mix anaerobic bioreactor is a conventional anaerobic bioreactor. Without coupling with the membrane filtration, the complete mix bioreactor may only be suitable for the solid or sludge digestion or for the small-scale wastewater treatment because it's low organic loading capacity could result in a large reactor volume for the treatment of a large wastewater flow, which becomes economically unfeasible. The UASB and EGSB can decouple the HRT from SRT through growing dense biomass to avoid the biomass wash-out under a short HRT condition, while the fluidized bed biofilm reactors through attached growth to retain the biomass in the bioreactor system. The current commercial high rate anaerobic bioreactor systems include Biothane UASB and EGSB, ADI hybrid bioreactor, BioPaQ$^{\delta}$ UASB, PAQ IC, etc. with the main market segments covering brewery, potato, pulp & paper, dairy, vegetable, etc. [6]. According to Kassam et al. [6], the current commercial high rate anaerobic reactors have successfully treated high strength wastewaters with COD up to 60,000 mg/L and achieved a COD removal higher than 85% in a HRT range less than 5 days. The biogas production rates of commercial full-scale anaerobic systems are usually around 500 L/kg COD. Integrating the high rate anaerobic bioreactors with the membrane filtration could further improve the effluent quality and operation stability.

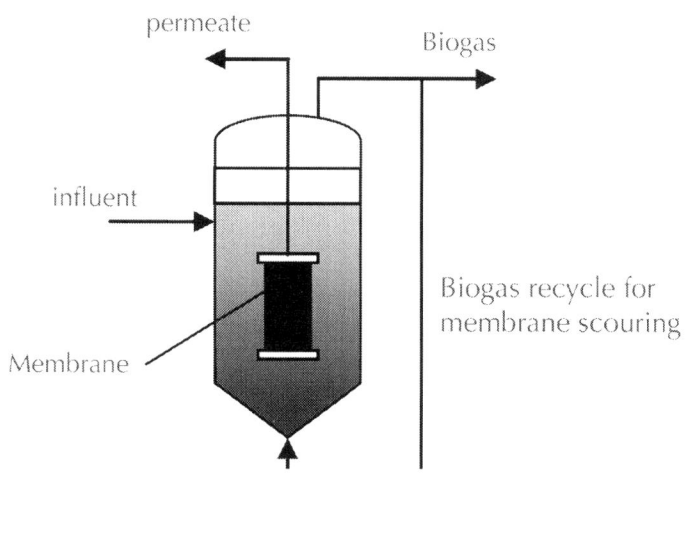

(a)

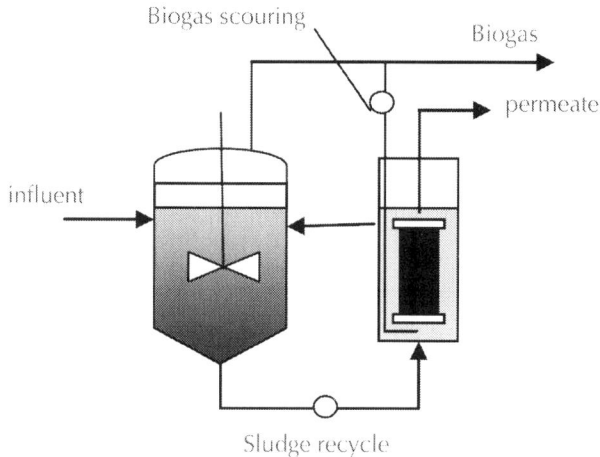

(b)

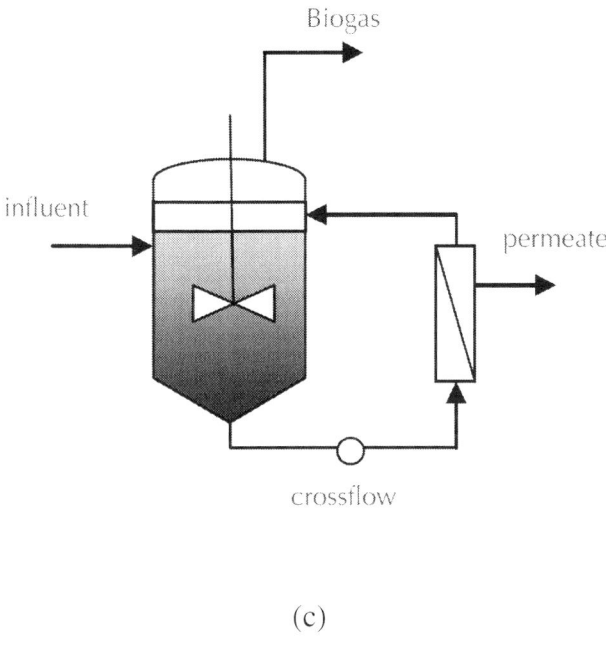

(c)

Figure 1: Different AnMBR system configurations. (a) Submerged membrane AnMBR; (b) AnMBR with external submerged hollow fiber membrane; (c) AnMBR with external cross flow membrane.

One of the key components of an AnMBR system is the membrane filtration system. As shown in Figure 1, two different membrane filtration modules, the cross flow pressurised membrane modules and the submerged membrane filtration, can be used in AnMBRs: the external cross flow membrane filtration usually uses the conventional plate & frame or the cylindrical hollow fiber cartridge configurations. In such system, the liquid cross flow is used to generate the surface shear to control the membrane fouling. The permeate flow is driven by the cross flow generated pressure or obtained by pump suction. The immersed or submerged membrane modules have been widely used in the aerobic membrane processes. The immersed membrane modules include the submerged flat sheet membrane modules and the submerged hollow fibre membrane modules. Figure 2(a) shows a schematic of the submerged flat sheet membrane panels which are usually arranged at a gap of around 8 to 12 mm with gas injected into these gaps to prevent the sludge accumulation and membrane fouling. For a full scale module, up to 100 panels can be connected to a common manifold to form a filtration unit. Figure 2(b) shows a typical design of a submerged hollow fibre membrane module. The submerged hollow fibre membrane module consists of the module headers and fibre bundles. Most of the hollow fibre membranes used in MBRs are polyvinylidene fluoride (PVDF) hollow fibres with OD/ID range of 1 - 2/0.65 - 1 (mm/mm) [7] and the fibre bundle can be packed into a curtain or cylindrical configuration. The current commercial submerged hollow fiber membrane modules can achieve a packing density around 160 m^2/m^3 tank volume, which can provide a total production capacity of 800 m^3/day at an average day design flux of 22 $L/m^2/day$ [8].

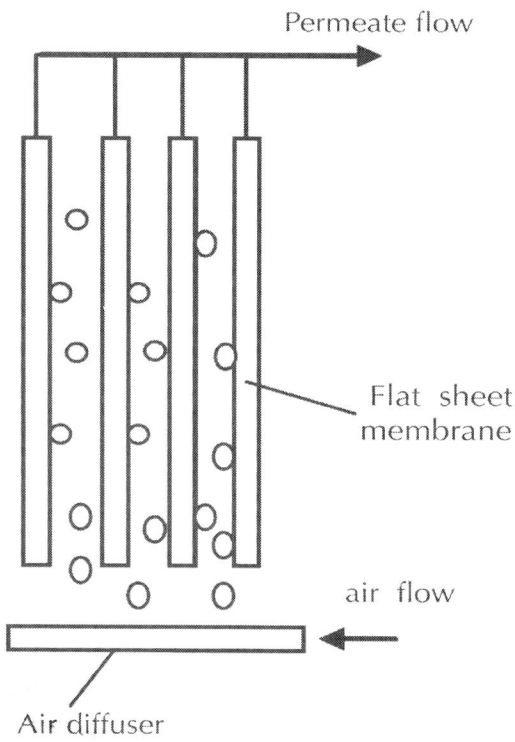

(a)

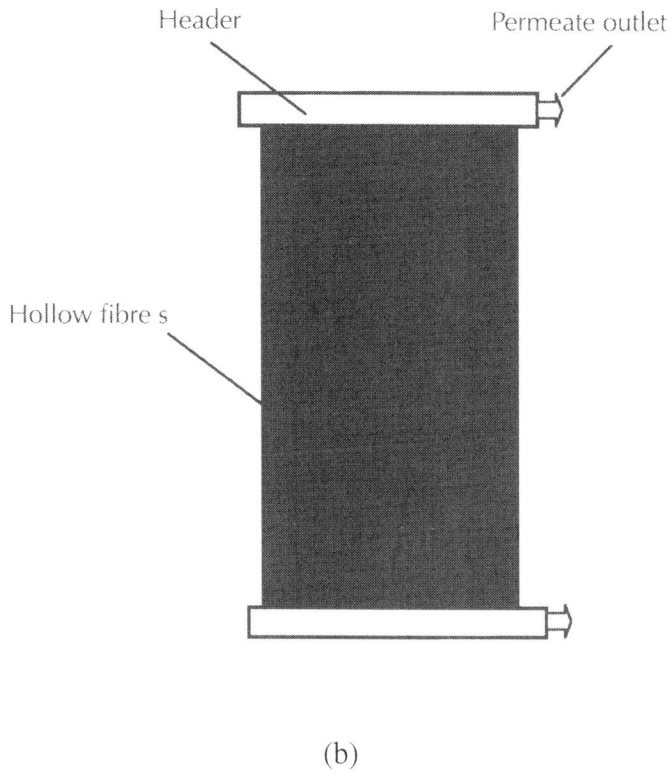

(b)

Figure 2: Schematics of submerged membrane modules. (a) Submerged flat sheet membrane modules; (b) Submerged d hollow fiber membrane modules.

APPLICATIONS OF ANMBRS IN WASTEWATER TREATMENT

AnMBRs have been tested for the treatment of a wide range of wastewaters and high solid content wastes, which include food processing wastewater, pulp and paper, landfill, municipal wastewater, etc. Table 1 summarises some reported results on the treatment of high strength wastewater of different sources by AnMBRs. In general, it has been showed that AnMBRs can achieve around 90% or higher COD removal and a methane production of 0.25 to 0.35 m^3 CH_4/kg COD. The organic loadings of AnMBRs ranged from 5 to 30 g COD/L/day, MLSS concentration from 15 to 30 g/L or higher, HRT from 1 to 25 days, and the membrane filtration flux from 5 to 10 LMH, were reported.

Recently, there is an increasing interest in the application of AnMBRs to municipal wastewater treatment [14-16]. Table 2 summarised some of the reported results on the applications of AnMBR in municipal wastewater treatment. Many studies showed that AnMBR can achieve efficient COD removal at temperature ranging from 20°C to 30°C in the treatment of municipal wastewater with a HRT from 24 to 6 hours tested. Most of studies showed that a long-term sustainable flux around 5 to 10 LMH was achievable for the municipal wastewater treatment AnMBRs. Lin et al. [14] conducted a feasibility evaluation of submerged anaerobic membrane bioreactor for municipal secondary wastewater treatment. A cost analysis based on their lab-scale test showed that the operational cost of an AnMBR could be only 1/3 of the aerobic treatment process and the energy generated from methane production can theoretically balance the energy required for the membrane biogas scouring [16]. Although it is feasible to treat municipal wastewater using AnMBR in terms of theoretical energy balance calculation, the full-scale commercial applications is still be limited by concerns of the stability of the treatment performance under ambient temperature; the effluent quality; and the energy recovery efficient under low influent COD conditions. The current experimental results showed that the temperature range for the effective anaerobic treatment of municipal wastewater is around 20°C to 30°C, so the anaerobic treatment of municipal wastewater may be still a challenge for places with cold winter since it is not economic feasible to heat large amount of wastewater flow. Martinez-Sosa reported that the methane production reduced from 0.27 L/gCOD to 0.24 L/gCOD when the temperature is reduced from 35°C to 20°C. Baek et al., [17], reported that no methane production was detected when treating municipal water with soluble COD ranging from 38 to 131 mg/L although 72% COD removal was observed. Dissolution of the methane in the treated effluent will also affect the energy recovery and increase the greenhouse gas emission from the discharged effluent.

Table 1: Treatment performances and process conditions of AnMBR used in high strength wastewater treatment

	Anderson et al. [9]	Choo & Lee [10]	Xie, et al. [11]	Van Zyl, et al. [12]	Zayen et al. [13]
Wastewater	Brewery	Distillery	Kraft evaporator condensate	Coal industria WW	Landfill
WW COD (g/L)	80 to 90	22.6	10	19.1	41
Temperature (°C)	35 to 37	53 to 55	36 to 38	37	37
OLR (kg COD/m³/day)	Above 30	1.5	22.5	Up to 25	6.27
HRT (day)	2.5 to 4.2	15	-	1.3	7
MLSS concentration (g/L)	Up to 51	-	8 to 12	36	-
COD removal (%)	99%	97%	93% to 99%	96.8%	90.7%
Gas production (m³ CH₄/kgCOD)	0.28	0.26	0.35	-	0.48

Table 2: Treatment performances and process conditions of AnMBR used in low strength municipal wastewater treatment

	Lew, et al. [14]	Martinez-Sosa [15]	Lin, et al. [16]
WW COD (g/L)	0.54	0.6*	0.342 - 0.527 (SCOD)
Temperature (°C)	25	20	30

Reactor type	Complete mix	Complete mix	UASB
Reactor volume (m³)	0.18	0.35	0.06
Membrane location	Side stream	Submerged	Submerged
module type	Hollow fiber (dead end)	Flat-sheet	Flat sheet
Membrane area (M²)	4	3.5	0.6
OLR (kg COD/m³/ day)	2.16*	0.4 - 0.9	1
HRT (day)	0.25	1.5 - 0.67	0.42
COD removal	88%	84 - 94	90
Gas yield (m³ CH₄/ kgCOD)	-	0.24	0.24
Membrane flux (LMH)	7.5	7	12

MEMBRANE FILTRATION PERFORMANCE IN ANMBR

One of the key design and operation parameters of the membrane filtration is the operation flux, which directly affects the capital and operation costs of AnMBRs. The membrane process design flux determines the membrane surface area or the number of the membrane modules required to treat a certain wastewater flow. The number of the membrane modules installed will further affect the size of the membrane tank, piping, and the consumption of the chemicals used for the membrane cleaning. The design flux of AnMBRs, at which a stable operation is supposed to be attained, should be determined based on relatively long term testing. Some of the main observations on the membrane filtration operation in AnMBRs are summarised as below:

- Fluxes ranging from 6.7 to 10 LMH were achievable for the treatment of the different wastewaters [18]. Based on the concept of the critical flux, controlling the operation flux is still the most critical strategies to achieve long-term stable operation.

- Intermittent permeation is important to achieve the long-term stable operation. The typical operation cycles could include 10 to 15 minutes permeation and 10 to 60 seconds relaxing. No obvious advantage of the membrane backwash over relaxing was observed.

- Many studies showed that the membrane filtration in AnMBRs could tolerate much higher suspended solid concentrations than the aerobic MBR systems, where the high MLSS concentration can considerably reduce the oxygen transfer efficiency, resulting in a drastic change in the filterability of the mixed liquor.

The membrane fouling is still the major factors limiting the efficiency of the AnMBR. For the sub-critical flux operation, rapid particle deposition on the membrane surface can be avoided and the membrane fouling is mainly caused by the graduate accumulation of colloidal or soluble SMP in the mixed liquor. Although many studies have showed that the SMPs mainly consist of protein and polysaccharides, little insights into the colloidal properties of the SMPs have been developed so far. Hence, the interaction mechanisms between the SMPs and the membrane surface are still unclear. In addition to the membrane fouling caused by the SMPs, the inorganic compounds can also play an important role in the membrane fouling in AnMBR. Herrera-Robledo et al. [19] reported that the cake layer was mainly composed of volatile solids (85%) and the rest were related to mineral matter, with the presence of inorganic salts containing Ca, Mg, Fe, P and Si. Studies showed that metal complexation could play an important role in the development of the irreversible membrane fouling. Lyko et al. [20] demonstrated that a significant amount of extracellular polymeric substances (EPS) could be released by treating the fouled membrane samples with Ca^{2+} selective cation exchange resin (CER). Iron has also been detected on the fouled membrane surface. Choo and Lee [10] suggested that the precipitation of the struvite ($MgNH_4PO_4 \cdot 6H_2O$) might play an important role in the inorganic membrane fouling in AnMBRs. The main strategies to control membrane fouling in AnMBR include the cross flow for the external membrane filtration and the biogas scouring for the submerged membrane filtration system. The membrane gas scouring technology has been well developed through the aerobic MBR development. The energy efficient membrane filtration gas scouring technologies include intermittent gas scouring [8] and, recently, the pulse gas scouring [21]. Other techniques tested for the membrane fouling control include using ultrasonic techniques [22], vibrating membranes [23], and adding chemicals or adsorbents, such as powder activated carbon, to improve the filterability of the mixed liquor or reduce the concentration of the soluble membrane foulants [24]. Recently, Kim et al. [5] tested directly immersing the hollow fiber membrane in a fluidized granular activated carbon (GAC) anaerobic biofilm reactor and reported that the fluidized GAC particle could be utilized for the membrane fouling control.

CONCLUSIONS

AnMBR concept was developed about three decades ago, but its commercial applications have been largely limited by the efficiency of the membrane filtration. Recent development in the large-scale wastewater treatment MBR has largely increased the potential of the anaerobic membrane technology as a practical, advanced full-scale wastewater treatment technology. The current research demonstrated that the AnMBR technology can be used for the treatment of a wide range of wastewaters with a great potential to recover energy and resources from both high strength wastewaters.

REFERENCES

1.   B. Q. Liao, J. T. Kraemer and D. M. Bagley, "Anaerobic Membrane Bioreactors: Applications and Research Directions," Critical Review in Environmental Science and Technology, Vol. 36, No. 6, 2006, pp. 489-530.http://dx.doi.

org/10.1080/10643380600678146

2.   D. Martinez-Sosa, B. Helmreich, T. Netter, S. Paris, F. Bischof and H. Horn, "Anaerobic Submerged Membrane Bioreactor (AnSMBR) for Municipal Wastewater Treatment under Mesophilic and Psychrophilic Temperature Conditions," Bioresource Technology, Vol. 102, No. 22, pp. 10377-10385. http://dx.doi.org/10.1016/j.biortech.2011.09.012

3.   M. L. Salazar-Pelaez, J. M. Morgan-Sagastume and A. Noyola, "Influence of Hydraulic Retention Time on Fouling in a UASB Coupled with an External Ultrafiltration Membrane Treating Synthetic Municipal Wastewater," Desalination, Vol. 277, No. 1, 2011, pp. 164-170. http://dx.doi.org/10.1016/j. biortech.2011.09.012

4.   L. B. Chu, F. L. Yang and F. L. Zhang, "Anaerobic Treatment of Domestic Wastewater in a Membrane-Coupled Expended Granular Sludge Bed (EGSB) Reactor under Moderate to Low Temperature," Process Biochemistry, Vol. 40, No. 3, 2005, pp. 1063-1070.http://dx.doi.org/10.1016/j.procbio.2004.03.010

5.   J. Kim, K. Kim, H. Young, H. Ye, E. Lee, C. Shin, P. Mccarty and J. Bae, "Anaerobic Fluidized Bed Membrane Bioreactor for Wastewater Treatment," Environmental Science & Technology Letters, Vol. 45, No. 17, 2011, pp. 576-581.

6.   Z. A. Kassam, L. Yerushalmi and S. Guiot, "A Market Study on the Anaerobic Wastewater Treatment Systems," Water, Air, and Soil Pollution, Vol. 143, No. 1-4, 2003, pp. 179-192. http://dx.doi.org/10.1023/A:1022807416773

7.   S. Chang, "Application of Submerged Hollow Fibre Membrane in Membrane Bioreactors: Filtration Principles, Operation, and Membrane Fouling," Desalination, Vol. 283, 2011, pp. 31-39. http://dx.doi.org/10.1016/j.desal.2011.03.025

8.   T. Bure and J. Cumin, "MBR Module Design and Operation," Desalination, Vol. 250, No. 3, 2010, pp. 1073- 1077. http://dx.doi.org/10.1016/j.desal.2009.09.111

9.   G. K. Anderson, B. Kasapgil and O. Ince, "Microbial Kinetics of a Membrane Anaerobic Reactor," Environmental

Technology, Vol. 17, No. 5, 1996, p. 449.http://dx.doi.org/10.1080/09593331708616407

10. K. H. Choo and C. H. Lee, "Membrane Fouling Mechanisms in the Membrane-Coupled Anaerobic Bioreactor," Water Research, Vol. 30, No. 8, 1996, pp. 1771-1780.http://dx.doi.org/10.1016/0043-1354(96)00053-X

11. K. Xie, H. J. Lin, B. Mahendran, D. M. Bagle, K. T. Leung, S. N. Liss and B. Q. Liao, "Performance and Fouling Characteristics of a Subme Rged Anaerobic Membrane Bioreactor for Kraft Evaporator Condensate Treatment," Environmental Technology, Vol. 31, No. 5, 2010, pp. 511-521. http://dx.doi.org/10.1080/09593330903527898

12. P. J. Van Zyl, M. C. Wentzel, G. A. Ekama and K. J. Riedel, "Design and Start-Up of a High Rate Anaerobic Membrane Bioreactor for the Treatment of a Low pH, High Strength, Dissolved Organic Waste Water," Water Science Technology, Vol. 57, No. 2, 2008, pp. 291-295. http://dx.doi.org/10.2166/wst.2008.083

13. A. Zayen, S. Mnif, F. Aloui, F. Fki, S. Loukil, M. Bouaziz and S. Sayadi, "Anaerobic Membrane Bioreactor for the Treatment of Leachates from Jebel Chakir Discharge in Tunisia," Journal of Hazardous Materials, Vol. 177, No. 1-3, 2010, pp. 918-923.http://dx.doi.org/10.1016/j.jhazmat.2010.01.004

14. B. Lew, S. Tarre, M. Beliavski, C. Dosoretz and M. Green, "Anaerobic Membrane Bioreactor (AnMBR) for Domestic Wastewater Treatment," Desalination, Vol. 243, No. 1-3, 2009, pp. 251-257. http://dx.doi.org/10.1016/j.desal.2008.04.027

15. D. Martinez-Sosa, B. Helmreich, T. Netter, S. Paris and F. bischof, "Anaerobic Submerged Membrane Bioreactor (AnSMBR) for Municipal Wastewater Treatment under Mesophilic and Psychrophilic Temperature Conditions," Bioresource Technology, Vol. 102, No. 22, 2011, pp. 10377-10385. http://dx.doi.org/10.1016/j.biortech.2011.09.012

16. H. Lin, J. Chen, F. Wang, L. Ding and H. Hong, "Feasibility Evaluation of Submerged Anaerobic Membrane Bioreactor for

Municipal Secondary Wastewater Treatment," Desalination, Vol. 280, 2011, pp. 120-126.http://dx.doi.org/10.1016/j. desal.2011.06.058

17. S. H. Baek, K. R. Pagilla and H. J. Kim, "Lab-Scale Study of an Anaerobic Membrane Bioreactor (AnMBR) for Dilute Municipal Wastewater Treatment," Biotechnology and Bioprocess Engineering, Vol. 15, No. 4, 2010, pp. 704-708. http://dx.doi.org/10.1007/s12257-009-0194-9

18. S. Kirmura, "Japans Aqua Renaissance'90 Project," Water Science & Technology, Vol. 23, No. 7-9, 1991, pp. 1573-1592.

19. H. Herrera-Robledo, J. M. Morgan-Sagastume and A. Noyola, "Biofouling and Pollutant Removal during LongTerm Operation of an Anaerobic Membrane Bioreactor Treating Municipal Wastewater," Biofouling, Vol. 26, No. 1, 2010, pp. 23-30.http://dx.doi.org/10.1080/08927010903243923

20. S. Lyko, D. Al-Halbouni, T. Wintgens, A. Janot, J. Hollender, W. Dott and T. Melin, "Polymeric Compounds in Activated Sludge Supernatant—Characterisation and Retention Mechanisms at a Full-Scale Municipal Membrane Bioreactor," Water Research, Vol. 41, No. 17, 2007, pp. 3894-3902. http:// dx.doi.org/10.1016/j.watres.2007.06.012

21. M. Kondo, J. Cumin, Y. Hong, R. Bayly, M. Gao and R. Rubin, "Reexamination of the Gas Sparging Mechanism for Membrane Fouling Control," Proceedings of the Water Environment Federation, WEFTEC 2010, New Orleans, pp. 6986-7007.

22. P. Sui, X. Wen and X. Huang, "Feasibility of Employing Ultrasound for On-Line Membrane, Fouling Control in an Anaerobic Membrane Bioreactor," Desalination, Vol. 219, No. 1, 2008, pp. 203-213. http://dx.doi.org/10.1016/j. desal.2007.02.034

23. A. Kola, Y. Ye, A. Ho, P. Le-Clech and V. Chen, "Application of Low Frequency Transverse Vibration on Fouling Limitation in Submerged Hollow Fibre Membranes," Journal of Membrane

Science, Vol. 409-410, 2012, pp. 54-65.http://dx.doi.org/10.1016/j.memsci.2012.03.017

24. H. Park, K. H. Choo and C. H. Lee, "Flux Enhancement with Powder Activated Carbon Addition in the Membrane Anaerobic Bioreactor," Separation Science and Technology, Vol. 34, No. 14, 1999, pp. 2781-2792. http://dx.doi.org/10.1081/SS-100100804

A Scenario Analysis of Future Energy Systems based on an Energy Flow Model Represented as Functionals of Technology Options

Yasunori Kikuchi[a, b], Seiichiro Kimura[b], Yoshitaka Okamoto[c], and Michihisa Koyama[b, c, d]

[a]Presidential Endowed Chair for "Platinum Society", The University of Tokyo, Ito International Research Center, 7-3-1 Hongo, Bunkyo-ku, Tokyo 113-0033, Japan

[b]International Institute for Carbon-Neutral Energy Research, Kyusyu University, 744 Motooka Nishi-ku, Fukuoka 819-0395, Japan

[c]Department of Hydrogen Energy Systems, Kyushu University, 744 Motooka Nishi-ku, Fukuoka 819-0395, Japan

ᵈInamori Frontier Research Center, Kyushu University, 744 Motooka Nishi-ku, Fukuoka 819-0395, Japan

ABSTRACT

The design of energy systems has become an issue all over the world. A single optimal system cannot be suggested because the availability of infrastructure and resources and the acceptability of the system should be discussed locally, involving all related stakeholders in the energy system. In particular, researchers and engineers of technologies related to energy systems should be able to perform the forecasting and roadmapping of future energy systems and indicate quantitative results of scenario analyses. We report an energy flow model developed for analysing scenarios of future Japanese energy systems implementing a variety of feasible technology options. The model was modularized and represented as functionals of appropriate technology options, which enables the aggregation and disaggregation of energy systems by defining functionals for single technologies, packages integrating multi-technologies, and mini-systems such as regions implementing industrial symbiosis. Based on the model, the combinations of technologies on both energy supply and demand sides can be addressed considering not only the societal scenarios such as resource prices, economic growth and population change but also the technical scenarios including the development and penetration of energy-related technologies such as distributed solid oxide fuel cells in residential sectors and new-generation vehicles, and the replacement and shift of current technologies such as heat pumps for air conditioning and centralized power generation. The developed model consists of two main modules; namely, a power generation dispatching module for the Japanese electricity grid and a demand-side energy flow module based on a sectorial energy balance table. Both modules are divided and implemented as submodules represented as functionals of supply- and demand-side technology options. Using the developed model, three case studies were performed. Required data were collected through workshops involving researchers and engineers in the energy technology field in Japan. The functionals of technologies were defined on the basis of the availability of data and understanding of the current and future energy systems. Through case studies, it was demonstrated that the potential of energy technologies can be analysed by the developed model considering the mutual relationships of technologies. The contribution of technologies to, e.g., the reduction in greenhouse gas emissions should be carefully examined by quantitative analyses of interdependencies of the technology options.

INTRODUCTION

Since the devastating East Japan earthquake on March 11, 2011, Japan has faced a serious short-term power shortage, as well as a large uncertainty in envisioning future energy systems towards the attainment of a low-carbon society [1]. A foresight approach will be useful for constructing a Japanese future vision and pathways to a sustainable energy society from this uncertain situation. Foresight approaches can be classified into forecasts, exploratory scenarios, technical scenarios, visions, backcasts and roadmaps [2]. Energy system design and analysis has become one of the largest issues, including not only energy supply but also demand [1]. Foresight by several methods, e.g., forecasts, visions or backcasts, has addressed the planning of energy-related systems consisting of various technologies [2]. Scenario analysis is one of the scientific approaches striving for possible and desirable systems based on specific objective functions. A technology assessment applies scenario analyses for visualizing the characteristics of available technologies (e.g., [3]) with a graphical representation method (e.g., [4]). Power supply systems with a target technology were analysed on the basis of the scenarios of market penetration [5]. A single optimal system cannot be suggested because the availability of infrastructure and resources, and the acceptability of the system should be discussed locally involving all related stakeholders in the energy system. While modelling and analysis for specific sectors in a country have been conducted, e.g., [6], the design of the system has a wide range of alternative candidates because of the local availability of resources [7], infrastructures [8] or motivation [9].

Energy systems can be represented as the combination of domestic energy flows, conversions and consumptions associated with processes in the industrial and civilian sectors, as well as inputs and outputs of materials across the boundary. To discuss a feasible energy system for Japan, it is necessary to consider the life cycles of both existing and emerging technology options with analysis of the temporal variations of energy usage, and the change of social and economic situations. Regarding energy systems, multiple stakeholders are involved and should be able to contribute to the construction and maintenance of a "benign" system [10] for sustainability. In particular, researchers and engineers of technologies related to energy systems should be able to perform the forecasting and roadmapping of future energy systems and indicate quantitative results of scenario analyses. At the same time, the prediction of future energy systems can facilitate motivation for technology development. Technical scenarios [2] on possible energy systems may strongly encourage technology developers to seek viable and effective solutions for energy. Roadmaps of technologies from the viewpoints of policymakers [11], researchers/engineers [12] or the citation network of journal papers [13] containing the potential of technologies both in operation and under research and development (R&D) can also be considered as references for characterizing the performance of technologies in the future. In this regard, however, such roadmaps should be carefully assessed to avoid overestimation of technologies. Energy technologies have various conditions, such as competitiveness or premises. To assess the roadmapping by researchers and engineers, all related developers should be able to access the rationales of roadmapping and assessment results. For such involvement of researchers and engineers, a quantitative model is required and should be shared in forms understandable by the users.

Various models of energy systems have been developed. The Long-range Energy Alternatives Planning System (LEAP) is a tool for energy policy analysis and climate change mitigation assessment [14]. The Asia–Pacific integrated model (AIM) has been utilized for analysing global and regional emission scenarios [15]. Based on the elasticity of technology penetration of an economic system, the market allocation of technologies was modelled [16] and analysed for the installation of specific technologies [17]. These existing models are more or less based on the economic optimization mechanism for analysing energy systems under the constraints on the maturity level of technologies such as fuel conversion efficiencies or performance factors. The analysis tends to focus on foresight into the penetration of specific technologies. While the technical information on future feasible technologies was estimated [18], the range of the attainable regions because of technology development has not been discussed in detail. For scenario-based analysis of technology R&D, a model starting from the design of micro-level parameters, e.g., device or operation parameters of energy technology, is needed. Each model has been adopted for the foresight analysis of energy systems depending on the objectives. Although some of them were applied to systems analysis for policy-making [19], no examples exist for the R&D of individual technologies except for technology assessment, where systems consisting of technologies have been analysed; e.g., district heating [20] or the material requirements for vehicle technologies [21]. To visualize the characteristics of technologies related to energy systems in detail and to establish a strong pipeline between researchers or engineers of energy technologies and other stakeholders, e.g., policymakers and the general public, a tool can be employed as a communication method that enables visualization of the potential of technologies developed in the future with quantitative analysis based on a logically developed mathematical model. Because energy systems have multiple scales, levels and aspects, the model should be able to modularize the system into subsystems understandable for stakeholders in the energy system. The scope of subsystems addresses not only energy demand sectors but also the functionals of energy technologies fulfilling heating/ cooling or lighting.

In this paper, we propose a procedure for model-based scenario analysis of energy systems into the future. In this analysis, the effects of energy flows because of the installation or development of technology options are shown by an energy flow model represented as functionals of technology options (EFM-FTO). To develop EFM-FTO, existing technology options are reviewed based on their functionals. Case studies of scenario analysis based on EFM-FTO are conducted for energy systems in Japan.

MATERIAL AND METHODS

Overall Framework of Scenario Analysis

The overview of EFM-FTO is shown in Fig. 1. EFM-FTO converts input societal and technical scenarios into a future energy flow in Japan by matching required and affordable services. Based on the input scenarios, societal parameters are denoted to specify the required services. Required service represents the demand for energy, not the consumption of fuel. The net efficiency of energy technologies is taken into account. The functional of affordable services is provided by a module of final service interactively connected with a module of centralized energy conversion by dispatching power sources based on electricity demand/supply. These two modules contain the subfunctionals considering technology packages. We established, mapped and linked submodels of feasible energy technologies not only for energy generation and supply technologies, e.g., photovoltaic (PV) power, wind power and new thermal power generation such as natural gas combined cycles and secondary batteries, but also for energy consuming technologies, e.g., home electric appliances and next generation vehicles. Through conversion of input technical scenarios into technological parameters, these modules can output affordable services and fuel consumption. Finally, EFM-FTO shows all of the results as a future energy flow based on societal and technical scenarios.

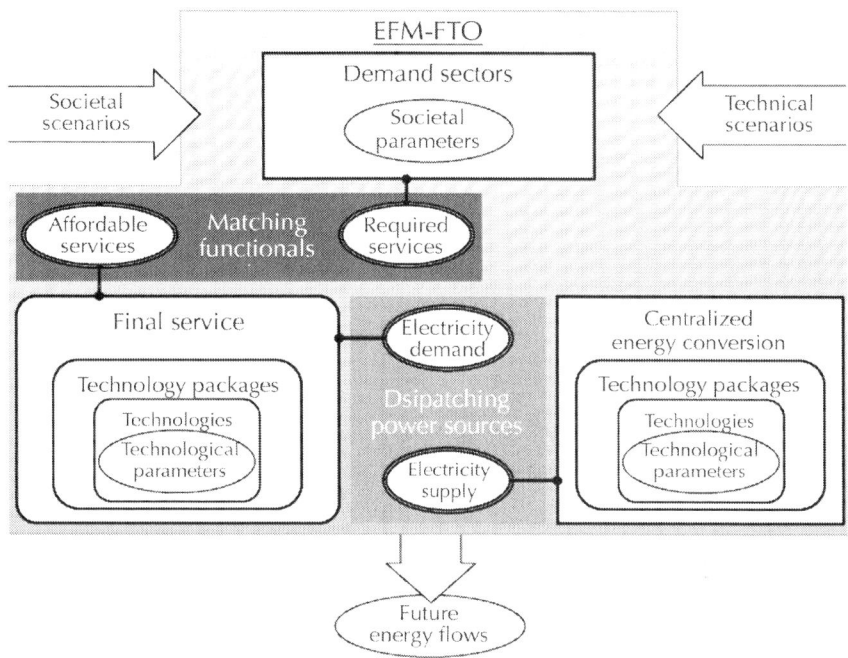

Figure 1: Overall framework of scenario analysis applying the EFM-FTO developed mathematical energy system model in this study.

Superstructure of the Usage of EFM-FTO

The superstructure of the usage of EFM-FTO is shown in Fig. 2. Through the usage of EFM-FTO, the model user can involve technology experts. The usage stage can be divided into three stages: analysis of the current system, alternative generation and simulation, where the alternative candidate systems consisting of technology options are designed and their performances and behaviours are simulated, and analysis of alternative systems. The first stage, analysis of the current system, shows the detailed performance and behaviour of the current energy system as conducted in a previous review [22]. In the second stage, the alternatives for energy technologies in final service and centralized power conversion are generated by interactive communication using EFM-FTO, which can be employed for visualization of the effects of alternative energy systems. The third stage conducts the interpretation for discussing future energy systems. Note that there may be feedbacks and iterations occurring at all intra- and inter stages to perform reanalysis and regeneration of alternatives, although they are not shown in Fig. 2.

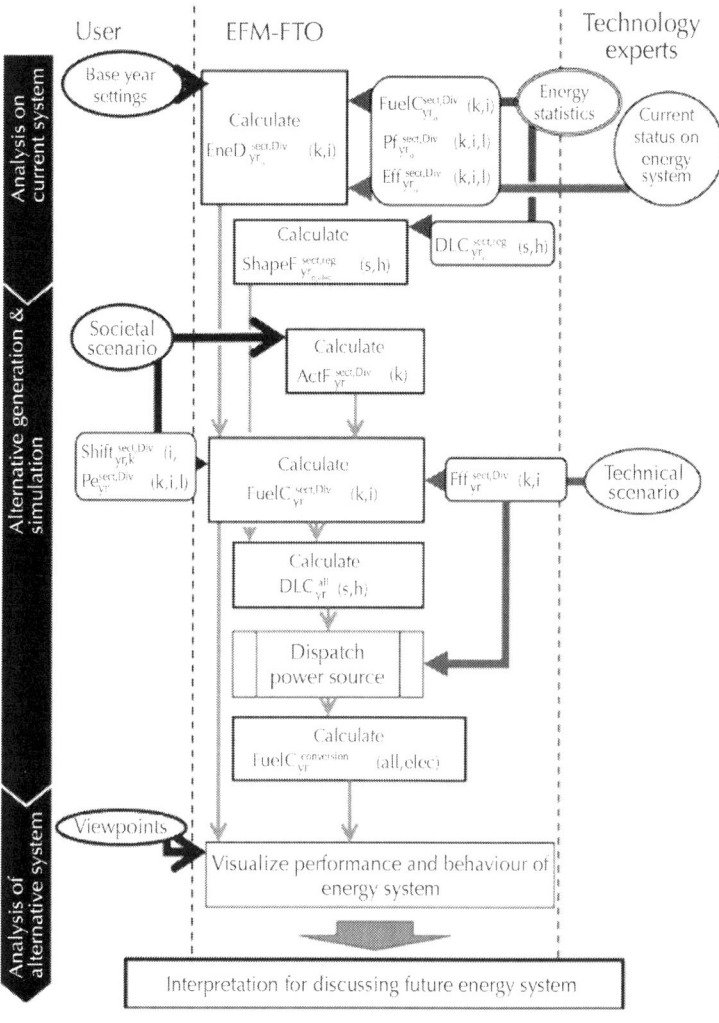

Figure 2: Procedure flow chart of energy technical scenario analysis applying the proposed EFM-FTO.

In the analysis of the current system, the detailed energy demand considers service, fuel, sector, division, subdivision and technology. The types of services are, for example, the heating or cooling in a house in the residential sector. The types of fuels were obtained from the list of fuels in Japanese statistics [23]. Divisions and subdivisions were defined as shown in Table 1. Technologies are discussed in the next section. In the first stage, $EneD_{yr_0}^{sect.Div}(k,i)$ (J-service/yr) is the energy demand [J] for service k by fuel i in division Div of sector $sect$ at reference year yr_0. Eq. (1) shows the definition of energy demand achieved by fuel consumption at the reference year.

$$EneD_{yr}^{sect.Div}(k,i) = FuelC_{yr}^{sect.Div}(k,i)$$
$$\cdot \sum_l \left(Pf_{yr}^{Div}(k,i,l) \cdot Eff_{yr}^{Div}(k,i,l) \right).$$
(1)

$$\text{where } \sum_l Pf_{yr}^{sect.Div}(k,i,l) = 1.$$
(2)

Table 1: The number of categories in the existing statistical data and EFM-FTO (see also Table S1)

Sector	Comprehensive energy statistics [23]	Greenhouse gas (GHG) inventory [24]	EFM-FTO
Residential	10 Divisions (Region)	10 Divisions (Region)	10 Divisions
			5 Usage
Commercial	9 Divisions	8 Divisions	8 Divisions
	15 Subdivisions		13 Subdivisions
			5 Usage

Industry	14 Divisions	14 Divisions	13 Divisions
	46 Subdivisions	15 Subdivisions	2 Subdivisions
	92 Smallest subdivisions		2 Usage
Transportation	2 Divisions	2 Divisions	14 subdivisions
	9 Subdivisions	9 Subdivisions	1 Usage
	14 Smallest subdivisions	14 Smallest subdivisions	
(Non-energy)	3 Divisions	2 Divisions	(Not considered)
	13 Subdivisions	6 Subdivisions	
		11 Smallest subdivisions	

In Eq. (1), fuel consumption, $FuelC_{yr}^{sect,Div}(k,i)$ (J-fuel/yr), is converted into energy demand by the technology penetration, $Pf_{yr}^{sect,Div}(k,i,l)$ (−) and the conversion efficiency of technology l, $Eff_{yr}^{sect,Div}(k,i,l)$ (J-service/J-fuel). Such parameters can be obtained from technology experts for the current status of the energy system with existing statistics [22]. Other required base year settings indicate which year is set as the reference year, together with other social information such as population or gross domestic product (GDP). In addition to discussion of the amount of consumed energy (J), the rate of energy supply, especially electricity supply, (W), is analysed in this stage. Consideration of the electricity load curve is important for addressing the electricity demand/supply structure [25]. As discussed by Koyama et al., most of the daily load curve of Japanese electric power is available from electric power companies [22], [26], [27], [28], [29], [30], [31], [32], [33] and [34]. Based

on the daily load curve at the reference year, $DLC_{yr_0}^{all}(s,h)$ (W), and other related statistics on final energy demand, the shape factor of

each electricity demand, $ShapeF_{yr_0,elec}^{sect,reg}(s,h)$, can be calculated. These parameters were defined as hourly parameters for 10 representative seasons, s; i.e., weekday or weekend of spring, summer, autumn and winter with peak days in summer and winter. $DLC_{yr_0}^{all}(s,h)$, the whole daily load curve, was calculated by summing the daily load curve of the 10 electric power companies.

$$DLC_{yr_0}^{all}(s,h) = \sum_{reg} DLC_{yr_0}^{reg}(s,h),$$

(3)

Where reg means the 10 regions in Japan. $DLC_{yr_0}^{reg}(s,h)$ is the average curve of 10 representative seasons in region reg. The shape factors were specified based on the calculated total daily load curve as follows.

$$ShapeF_{yr_0,elec}^{sect,reg}(s,h) = ShapeAF_{elec}^{sect}(s,h) \cdot \frac{DLC_{yr_0}^{reg}(s,h)}{\max_{s,h} DLC_{yr_0}^{reg}(s,h)},$$

(4)

Where $ShapeAF_{elec}^{reg}(s,h)$ is the allocation factor of the ratio of electricity demand in sector sect (–), and Eq. (5) holds. Such allocation factors were extracted from reports by Japanese ministries (e.g., [35]).

$$\sum_{sect} ShapeAF_{elec}^{sect}(s,h) = 1.$$

(5)

As well as $ShapeF_{yr_0,elec}^{sect,reg}(s,h)$, the power consumption of sector sect in region reg is needed to estimate daily load curves of the sectors. The monthly power consumption of sectors in a specific region, $FuelC_{yr_0^m}^{sect,reg}(elec)$ (W h/month), is available as public reports from the federation of electric power companies of Japan [36]. Using these data, the daily power consumption of sector sect in region reg during season s was estimated using Eq. (6).

$$\text{FuelC}^{sect,reg}_{yr_0^{day},s^{(WD \text{ or } WE)}}(\text{elec})$$

$$= \frac{\dfrac{\text{Days}_{s,WD \text{ or } WE} \cdot \sum_h \text{DLC}^{reg}_{yr_0}(s^{(WD \text{ or } WE)},h)}{\text{Days}_{s,WD} \cdot \sum_h \text{DLC}^{reg}_{yr_0}(s^{WD},h) + \text{Days}_{s,WE} \cdot \sum_h \text{DLC}^{reg}_{yr_0}(s^{WE},h)} \cdot \sum_{m_s} \text{FuelC}^{sect,reg}_{yr_0^{m_s}}(\text{elec})}{\text{Days}_{s(WD \text{ or } WE)}} \cdot$$

$$(6)$$

Where Days_s is the number of days in season $s(d)$, s^{WD}/s^{WE} represents the weekday/weekend in a particular season, and m_s is the number of months included in each of the four seasons; i.e., spring includes April and May, summer is June to September, Autumn is October and November, and winter is December to March. Because the available statistical data, $s^{(WD \text{ or } WE)}$, include the power consumption on both weekdays and weekends in a month, they are divided into the daily power consumption during weekdays and weekends (Wh/day); i.e., $\text{FuelC}^{sect,reg}_{yr_0^{day},s^{WD}}(\text{elec})$ and $\text{FuelC}^{sect,reg}_{yr_0^{day},s^{WE}}(\text{elec})$ in Eq. (6). Based on these estimated values, the daily load curves of sector *sect* in region *reg* were analysed using Eq. (7).

$$\text{DLC}^{sect,reg}_{yr_0}(s,h) = \frac{\text{ShapeF}^{sect,reg}_{yr_0,elec}(s,h)}{\sum_h \text{ShapeF}^{sect,reg}_{yr_0,elec}(s,h)} \cdot \text{FuelC}^{sect,reg}_{yr_0^{day},s}(\text{elec}).$$

$$(7)$$

Through analysis of the current system with EFM-FTO, key performances and behaviours of the current energy system can be analysed and shown based on energy flows. In the next stage in Fig. 2, alternative generation and simulation, societal and technical scenarios are assumed, and the change in such key performances or behaviours are analysed. Activity factors of energy consumed per service k in division *Div* of sector sect, $\text{ActF}^{sect,Div}_{yr}(k)$, was defined in this paper. These factors are set to address the energy intensity of the change of society. The base year activity is set as 1; i.e., $\text{ActF}^{sect,Div}_{yr_0}(k) = 1$ If the $\text{ActF}^{sect,Div}_{yr}(k)$ is increased, the energy demand in division *Div* of sector *sect* is increased. This factor can be affected by various societal scenarios, including population, the change of the economic activities in Japan and industry structure. EFM-FTO supports the implementation of such aspects as factors connected with the final energy demand in year yr. As described in Eq. (8), the fuel consumption is calculated based on the maturity of technology efficiency, market penetration of the technology, fuel shift in demand sectors and activity factors.

$$
\begin{aligned}
\text{FuelC}_{yr}^{sect.Div}(k.i) = {}& \text{ActF}_{yr}^{sect.Div}(k) \cdot (\text{EneD}_{yr_0}^{sect.Div}(k.i) \\
& + \sum_j \text{Shift}_{yr.k}^{sect.Div}(i.j) \cdot \text{EneD}_{yr_0}^{sect.Div}(k.j)) \\
& \cdot \sum_l \frac{Pe_{yr}^{sect.Div}(k.i.l)}{\text{Eff}_{yr}^{sect.Div}(k.i.l)} \cdot
\end{aligned}
\tag{8}
$$

$$
\text{where } \sum_l Pe_{yr}^{sect.Div}(k,i,l) = 1.
\tag{9}
$$

The market penetration of technology l for service k by fuel i in division Div of sector $sect$, $Pe_{yr}^{sect,Div}(k,i,l)$, has a similar definition to $Pf_{yr}^{sect,Div}(k,i,l)$ in Eq. (1). $Pe_{yr}^{sect,Div}(k,i,l)$ Provides the ratio of applied technologies within a certain service k achieved by energy, while $Pf_{yr}^{sect,Div}(k,i,l)$ is the ratio of technologies consuming a certain fuel.

$Shift_{yr,k}^{sect,Div}(i,j)$ And $Eff_{yr}^{sect,Div}(k,i,l)$ are the main factors of societal and technical scenarios, respectively. These parameters are internally connected to each other, because the technologies have strong connections for providing a service to final users. In the construction of models based on Eq. (8), the functionals of technologies become embedded structures representing the relationships of technologies in final energy use.

In addition, especially in electricity demand, the daily load curve is important for designing future power grids considering their load-following capacity. As shown in Eq. (7), $\text{FuelC}^{sect,reg}_{yr^{day},s}(\text{elec})$ is required for estimation of the future daily load curve. It was estimated based on the simulated fuel consumption and current load curve described in Eq. (10).

$$\text{FuelC}^{sect,reg}_{yr^{day},s}(\text{elec}) = \text{FuelC}^{sect,reg}_{yr_0^{day},s}(\text{elec})$$

$$\cdot \frac{\sum_{Div}\sum_k \text{FuelC}^{sect,Div}_{yr}(k, i = \text{elec})}{\sum_{Div}\sum_k \text{FuelC}^{sect,Div}_{yr_0}(k, i = \text{elec})}. \tag{10}$$

After obtaining $\text{FuelC}^{sect,reg}_{yr^{day},s}(\text{elec})$, the same algorithm shown in Eq. (10) can be applied to estimate future daily load curves. The daily load curve is strongly dependent on seasonal demands, such as air conditioning. Eq. (11) gives the shape factors considering societal and technical scenarios. Such seasonal functionals of energy technologies are addressed in this equation. If the functionals have season- and time-specific demands represented as s' and h' in Eq. (11), their factors are calculated considering the specificities.

$$\text{ShapeF}^{sect,Div,k'}_{yr,elec}(s', h') = \text{ShapeF}^{sect,Div,k'}_{yr_0,elec}(s', h')$$

$$\cdot \frac{\text{FuelC}^{sect,Div}_{yr}(k, i = \text{elec})}{\text{FuelC}^{sect,Div}_{yr_0}(k, i = \text{elec})}. \tag{11}$$

Using these factors, $\text{DLC}^{all}_{yr}(s,h)$ can be estimated using Eq. (7).

Based on the estimated $\text{DLC}^{all}_{yr}(s,h)$, the composition of the power sources was specified by the module dispatching power sources to meet the demand, as shown in Fig. 1 and Fig. 2. In EFM-FTO, an existing dispatching mechanism [22] was adopted for this module. This mechanism was developed for analysing centralized power systems on the basis of technology characteristics. Each power generation technology has its own characteristics on the initial and running costs, load-following capacity or securing of resources. The existing mechanism enables users to consider these characteristics in meeting the demand of seven representative days: weekday/weekend of summer, winter and midterm seasons, and the peak day in summer. In this regard, there are an infinite number of possibilities for power-supply systems to meet the demand. Therefore, the mechanism applied an optimizing approach to specify a power-supply system for certain conditions of DLC under defined constraints. The objective function in this optimization is the total cost attributable to centralized power companies; i.e., the aggregation of the existing 10 electric power companies in Japan. The minimization of the cost is the direction of optimization here. The cost includes construction and running costs of centralized power generation plants. The main parameters in the power supply system are the installation capacities of technologies and the hourly operation of the installed technologies. They have constraints on the upper/lower limitations, and requirements to meet the DLC take into account the load-following capacities of the technologies. As the constraints on this optimization, the current capacities of technologies installed in the Japanese power supply in 2010 were obtained from the literature (e.g., [37]). The surplus power generation for energy security is also defined as a constraint in optimizing the parameters. The decommissioning of technologies is dependent on the legal durable years; i.e., 40 years for most power plants. The cost changes of fossil resources are also constraints, which were defined based on available statistical data [38] and other reports [39] and [40]. The installation of PV power generation and wind power turbines (WTs) is considered as one of the constraints. The power supplied from installed PV and WT is assumed to reduce DLC, where the operation ratio of PV and WT was carefully addressed considering the weather conditions in Japan. This algorithm supports alternative generation in the power generation system by minimizing cost and

calculates the $FuelC_{yr}^{conversion}$ (power supply, i). Existing literature or reports on renewable technologies (e.g., [41]) can be applicable for setting the total capacity and operation ratio of PV and WT. In dispatching power sources to meet the electricity demand, an existing optimization algorithm [42] was adopted, which has been applied for scenario analysis of future power sources [5].

In the third stage in Fig. 2, the generated alternative system was analysed from the viewpoint of energy flows. For example, greenhouse gas (GHG) mitigation can be one of the key performance indicators for future energy flow. It can be shown by using the

GHG emission inventories for $\mathrm{FuelC}_{yr}^{sect,Div}(k,i)$. The $\mathrm{DLC}_{yr}^{all}(s,h)$ can be an important behaviour of energy systems depending on the composition of the power sources. These indicators can be visualized by using EFM-FTO and support the interpretation for discussing future energy systems.

Selectable Technology Options

Table 2 shows the selectable technology options in the current version of EFM-FTO. EFM-FTO has a structure modularizing energy system based on final demand and energy conversion sectors. The model structure enables addition or improvement of functions

describing technologies by revising, for example, $\mathrm{Pe}_{yr}^{sect,Div}(k,i,l)$,

$\mathrm{Shift}_{yr,k}^{sect,Div}(i,j)$ or $\mathrm{Eff}_{yr}^{sect,Div}(k,i,j)$. The current version of EFM-FTO utilized in the case studies in the next section was developed through interviews and hearings, including workshops geared towards technology roadmapping with experts and researchers of energy technologies [12]. The societal and technical scenarios modelled in the current version of EFM-FTO are briefly introduced below (see also Table S1).

Table 2: Selectable technology options in the current version of EFM-FTO

		Technology option
Energy demand sector	Residential	Change in conversion efficiency from fuel to energy demand
		–Home electric appliances (Air conditioner for cooling or warming, lighting and refrigerator)
		–Hot water supply devices using fuel and electricity
		–Kitchen system
		Penetration of distributed power source (SOFC)
		Installation of solar-powered hot water system
		Shift in fuel
		–Electrification in kitchen using fuel gas
		–Electrification in hot water supply using fuel gas
		–Electrification in heating using kerosene
	Commercial	Change in conversion efficiency from fuel to energy demand
		–Home electric appliances (Air conditioner for cooling or warming, lighting and refrigerator)
		–Hot water supply devices using fuel and electricity
		–Kitchen system
		Installation of solar-powered hot water system
		Shift in fuel
		–Electrification in kitchen using fuel gas
		–Electrification in hot water supply using fuel gas
		–Electrification in heating using kerosene

	Industry	Installation of energy-saving technology
		−Membrane-distillation hybrid system
		−Utilization of low-temperature exhaust heat
		−Energy regeneration in crane or motor
		−Utilization of biomass in agriculture and forestry
		−Energy saving in steel plant (CCS and hydrogen production)
		−Energy management in small- and medium-sized enterprises
		Upgrading and improvement in private co-generation system
		−Chemically recuperated gas turbine
		−Gas engine
	Transportation	Improvement in fuel efficiency
		−Internal combustion engine vehicle (ICV) using gasoline, diesel and compressed gas
		−Train, ship and airplane
		Penetration of new-generation automobile technology
		−Internal combustion hybrid vehicle (ICHV) using gasoline, diesel and compressed gas
		−Fuel cell hybrid vehicle (FCHV)
		−Battery electric vehicle (BEV)
		Implementation of hydrogen production system for FCHV
		−Reformulation of fossil resource (kerosene or natural gas)
		−Electrolysis of water with or without load adjustment on the power grid

Energy conversion	Centralized conversion sector	Installation and improvement in fuel efficiency by replacing conventional power plants
		–Coal, oil and LNG fire, LNG gas combined cycle (LNG-GCC), nuclear, hydroelectric and geothermal
		Implementation of storage technology
		–Water pump, grid battery or vehicle to grid
		–Request of demand response
		–Operation of power supply technologies considering load-following capacity
	Distributed energy	Implementation of PV power generation and WT

Residential and Commercial Sectors

For the residential sector, $ActF_{yr}^{residential,Div}$ is strongly correlated with the population of each division (region, see also Table S1), while

$ActF_{yr}^{commercial,Div}$ is dependent on the division. For example, the real estate and house services may be affected by the change in population, although the retail trade may depend on the economic activities, which are not always decided by population. Based on this interpretation, can be defined individually for divisions. Regarding the fuel consumption to meet the energy demand, both residential and commercial sectors have similar alternative technologies, as shown in Table 2. They are mainly divided into the change in conversion efficiency from fuel to energy demand, additional installation of solar-powered hot water systems and the shift in fuel. For the residential sector, the penetration of solid-oxide fuel cells (SOFCs) as a distributed co-generation system is included. Note that the operation mode of SOFC in the residential sector is dominated by hot water supply. This means that the operation ratio of SOFC is decided by the demand for hot water. Electricity can be regarded as a by-product of hot water supply.

Industrial Sector

Because the service required in the industrial sector has a wide range of possibilities, heating/cooling and electric services were defined as services provided by energy. $ActF_{yr}^{commercial,Div}$ is strongly affected by the economic activities in Japan, which are not directly related with the change in population. The product life cycle is not always located within Japan, which means that some raw materials and final products are imported. Therefore, it was assumed that the activities of industries have no relationship with the settings of the life cycle of products or technologies installed in Japan for simplifying the calculation within EFM-FTO. Regarding technologies we adopted in the model, the installation of energy-saving technologies and co-generation systems are the main alternatives. In this regard, the unique technologies and systems were modelled in the current version of EFM-FTO to consider the individual conditions of divisions in the industrial sector. For example, the residues and by-products of agricultural processes can be applied as one of the renewable resources. Hydrogen as a by-product in steel plants can be utilized as an unused resource. In addition to such technologies, private co-generation plants can be upgraded to improve the heat/electricity balance of generation and utilization. These technologies may contribute to future energy system.

Transportation Sector

The services of the transportation sector are mainly divided into passenger and freight. $ActF_{yr}^{transportation,Div}$ is dependent on what is moved. In the statistics, fuel consumption is divided into passenger and freight. In the passenger component, home-use private cars may be affected by population considering the recent saturation of car ownership ratio. On the other hand, freight is based on economic activities. Regarding technologies, the improvements of fuel efficiency were modelled for vehicles, trains, ships and airplanes. Such improvement is achieved by, for example, the use of lightweight materials or friction control. Energy regeneration around brake systems is also a technology for improving fuel efficiency. In addition, the penetration of new-generation vehicle technologies will be one of the largest changes in the transportation sector from 2010 to 2050. The conventional internal combustion engine vehicles (ICEVs) and hybrid electric vehicles (HEVs) use gasoline, diesel and compressed gas as fuel. Fuel cell hybrid vehicles (FCVs) need hydrogen production and supply infrastructures. FCVs require large-scale hydrogen production, which necessitates hydrogen production technologies such as the reformulation of fossil resources, e.g., kerosene or natural gas, and the electrolysis of water with or without load adjustment on the power grid. Battery electric vehicles (BEVs) also require an infrastructure in society to charge batteries.

Energy Conversion Sectors

The requirements of the energy conversion sector are internally specified in EFM-FTO; i.e., $DLC_{yr}^{all}(s,h)$ to meet this demand, technologies are mixed considering the installation and improvement in fuel efficiency by replacing conventional power plants; e.g., coal, oil and LNG fire, LNG gas combined cycle (LNG-GCC), nuclear, hydrological and geothermal. The implementation of storage technologies, e.g., pump hydro, grid storage and vehicle to grid, is also modelled and can be considered. Regarding storage, demand response techniques can be a method for avoiding or reducing storage systems or increasing the operation ratio of installed plants. As distributed power connecting to the grid, PV and WT power generation are taken into account. To specify the installation and operation of energy conversion technologies, load following capacities of technology options are considered as adopted in existing power source modelling [42]. Note that 25.9 GW pump hydro has been installed in Japanese power grid as one of the storage, which can be employed to mitigate the frequency fluctuation of the grid caused by PV and WT.

CASE STUDIES

Objective

Based on the developed EFM-FTO, three case studies were performed. The first case study, named the Base Scenario Analysis, demonstrates the simulation by EFM-FTO. A base technical scenario was generated through interviews and hearings with technology experts, and investigation of the existing literature. One scenario was analysed by EFM-FTO, and several key performance indicators on the aspects of energy flows in Japan were evaluated and visualized. The second case study, named Technology Installation in the Residential Sector, demonstrates a scenario analysis of technology options for the residential sector. The effects of societal scenarios, e.g., population change, and technical scenarios, e.g., installation of technologies, on GHG emission in the residential sector are recognized. The effects of the order of penetration of technologies on the contribution of each technology to GHG emission reduction are quantified through scenario sensitivity analysis. The third case study, named Technology Installation in the Transportation Sector, conducted a technology assessment of vehicle-related technologies. Here, while the penetration of vehicle technologies was fixed at a certain level, their tank-to-wheel efficiencies were changed in specific ranges. The priorities of vehicle technologies under various conditions, including well-to-tank scenarios, were visualized. Through these three case studies, the usability of EFM-FTO is presented.

In the following analyses, the current status of Japanese energy flows in 2010, as investigated through the literature or existing databases [22], was set as the constraints. The accounting period in optimization for generating alternatives to the power grids was set from 2010 to 2050 in this study. The construction cost, depreciation conditions, fixed property tax, interest rate, limitation of operation ratio and load-following capacity were also set as constraints.

Scenario Settings and Simulation Results

Base Scenario Analysis

Settings

The scenario settings are shown in Tables S2–5. The commonly used settings for future forecasting of population and the activity factor are shown in Table S2. The activity factor was defined as the activity per capita increasing 1% per year. It led to an increase of activity factors by 2040. Because of the rapid decrease in the Japanese population [43], the activity factor was decreased from 2040 to 2050, as shown in Table S2, even though the activity per capita in 2050 increased about 48% compared with that in 2010. Technical scenarios on residential and commercial sectors are shown in Table S3. The elements of the technical scenarios are based on the selectable technologies in the current version of EFM-FTO listed in Table 2. The coefficient of performance (COP) of heat pumps is one of the important parameters and has been greatly improved [44]. Because COP is used as an absolute value in the electrification of heating using Eq. (8), its setting was initially analysed using a statistical method. The replacement of heat pumps in the market stock was estimated by the population balance analysis of home electric appliances based on the Weibull distribution function applied in material flow analysis of products (e.g., [45]). The trend of COP values based on the top runner approach was adopted as the COP of new air heat pumps replacing old air heat pumps. The average COP of the market stock was estimated by the replacement frequency and COP of new heat pumps. Regarding heat pumps in cold regions, the reduction ratios in COP because of the low air temperature and defrosting were estimated by existing external models developed by CRIEPI Japan [46], as discussed by Koyama et al. [22]. Other values in Table S3 were basically defined and set through interviews and hearings with experts in each technology

[12]. The shipment of SOFC for residential sectors was based on the trend of current installation records [47]. The improvement of the efficiency of home electric appliances such as refrigerators was estimated on the basis of existing data (e.g., [48]) and the latest technology reviews (e.g., [49] and [50]). The technical scenarios in the industrial sector are shown in Table S4. Existing discussion and technology roadmaps [12] were adopted here, and the energy-saving ratios were extracted. Each technology has target applicable divisions, as shown in Table S4. The reduction ratios were applied for the target divisions in EFM-FTO. Table S5 organizes the technical scenario settings in the transportation sector. Regarding vehicle technologies, a systematic survey and analysis was performed to specify the future possible range of fuel efficiencies. Table S6 organizes the adopted fuel efficiencies of vehicle technologies. The energy required for vehicle driving was set to increase gradually towards 2050 because of the following results indicating a weight increase in vehicles. Fig. S1 shows the trends of sales data of ordinary and lightweight minipassenger vehicles as a function of vehicle weights, as well as the future projection by logarithmic approximation based on their records [51] and [52]. For both types of passenger vehicles, the average weight has increased in the past 10 years, and this trend is expected to continue in the future. The number of vehicles owned per capita in Japan and the ratio of minivehicles out of all passenger vehicles are shown in Fig. S2. Both figures show a gradual increase, for which the future values are given by extrapolation of the approximate curve. Considering population projections for Japan [43], the number of ordinary passenger vehicles and minipassenger vehicles owned in the future was estimated as shown in Fig. S3. Finally, the weighted average of the vehicle weight was calculated as shown in Fig. S1, which has a link with the energy efficiency of vehicles. These results give a strong foundation for the expectation that the necessary energy for vehicle travel gradually increases. In the base scenario in this case study, the mean values were adopted. For airplanes, diesel and electric trains and ships, the fuel consumption was set as shown in Table S5. The fuel efficiencies are assumed to be improved by use of lightweight materials or friction control [12]. As with power

generation, the assumptions were the same as those addressed in the previous discussion on energy roadmapping (e.g., [12], [22] and [41]). The power generation by nuclear, coal, oil and LNG thermal plants were set as their installed capacities are gradually decreased due to the decommission after the durable years. The water hydro power plants supply excluding pumped water power plants has a constant installed capacity for the next four decades. The capacity of geothermal power generation is increased, which may be possible because of the promotion of its installation by feed-in tariff and deregulation in Japan [22]. Regarding PV and WT connected to the power grid, their installed capacities were extracted from existing roadmapping literature [41]. Their operation ratios were set on the basis of the current average operation ratio and an assumption that the stakeholders may give prior attention to implementing PV, resulting in the decrease of the operation ratio of PV with the increase in the installation of PV (see also Fig. S4).

Results

Based on the scenario parameters described in the previous section, a set of energy flows was calculated by EFM-FTO. The performance and behaviour of the specified energy flows are introduced below. Fig. 3 shows the total GHG emission in this case study. Fig. 3(a) shows the direct GHG emission where energy conversion is included as a sector, while the GHG emission from energy conversion is included in the final demand sectors in Fig. 3(b). The reduction in total GHG emission was 43.7% in 2050 from 2010. As also shown in Fig. 3, residential and commercial sectors generated indirect GHG emission by consuming electricity. In this regard, however, the ratio of indirect/direct GHG emission was decreased because of the energy saving and the significant improvement of GHG emission from grid power, as shown in Fig. 3(a). This is because the proportion of fire power plants was decreased in 2050 because of the installation of renewable power sources. Fig. 4(a) shows the installed capacity of energy conversion technologies. The ratio of renewable power sources in installed capacity increased from about 13% in 2010 to

74% in 2050. This tendency is the same in the power generation shown in Fig. 4(b), where the ratio of renewable power sources in power generation was increased from about 19% in 2010 to 61% in 2050. Because the installed renewable resources, i.e., hydro and geothermal, in 2010 have a higher operation ratio than those in 2050, i.e., WT and PV, the values in the ratios above increased from 13% in installed capacity to 19% in power generation in 2010, while they decreased from 74% to 64% in 2050. The daily load curves of peak days in summer and weekdays in winter in 2050 are shown in Fig. 4(c and d), respectively. EFM-FTO can show the composition of power sources on daily load curves. Both days applied pump hydroelectricity into the load-following sources. The large peak power supply by PV may not be accepted as power mix without backup power sources or large-scale storage technologies. The stable supply of electricity may not be as promised, which conflicts with the Electricity Business Act in Japan. Considering the expectation of available power supply from unstable power sources such as PV and WT, the composition of power sources should be discussed. Fig. 4(e and f) show the total resource use. It is demonstrated that the technical scenario can reduce the amount of fossil resources used instead of increasing the renewable resources. In addition to these results, detailed process behaviours are also shown. For example, Fig. 5 shows the composition of the demand of multi-products from oil refineries. The current oil refineries produce gasoline in the largest proportion in 2010. The balances of the product composition were gradually changed, and then LPG and diesel have relatively high proportions in 2050. This means that the process system of the oil refinery should be able to meet this demand on a long-term basis. Fig. 6 shows the relative changes in indicators; i.e., grid cost calculated in the power sources dispatching module, GHG intensities per W h-consumption, per capita and per GDP. Because the grid cost is utilized as an alternative generation of energy conversion and supply system, it has less meaning as an absolute value for the indicators used for decision-making. GHG intensity per W h-consumption was calculated using GHG emission from all activities associated with energy conversion and power supply and actual demand for electricity. The difference between electricity consumption and generation is mainly the loss in power supply, including the loss around substations, the loss from electric transmission and the margin of supply capacity. As shown in Fig. 6, the GHG intensity was significantly improved. This is also identified in Fig. 3. The GHG intensity per capita resulted in a decrease compared with 2010. Because of the reduction in population, much effort is required to decrease the GHG intensity per capita. This result shows that the adopted technical scenarios can effectively decrease GHG emission. Regarding GDP, it is assumed that Eq. (12) applies. Based on Eqs. (12) and (13) was developed.

$$\frac{ActF_{yr}}{GDP_{yr}} = Const. \tag{12}$$

$$\frac{ActF_{yr_0}}{GDP_{yr_0}} = \frac{ActF_{yr}}{GDP_{yr}} \quad \therefore GDP_{yr} = \frac{GDP_{yr_0}}{ActF_{yr_0}} \cdot ActF_{yr}. \tag{13}$$

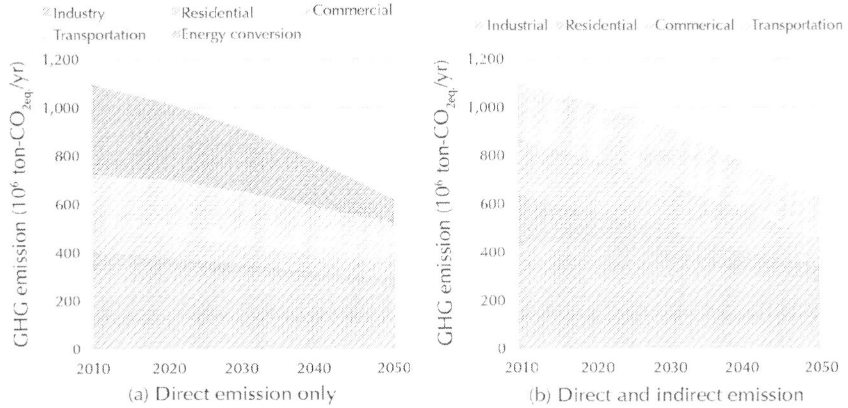

Figure 3: Total GHG emission in the set of base societal and technical scenarios.

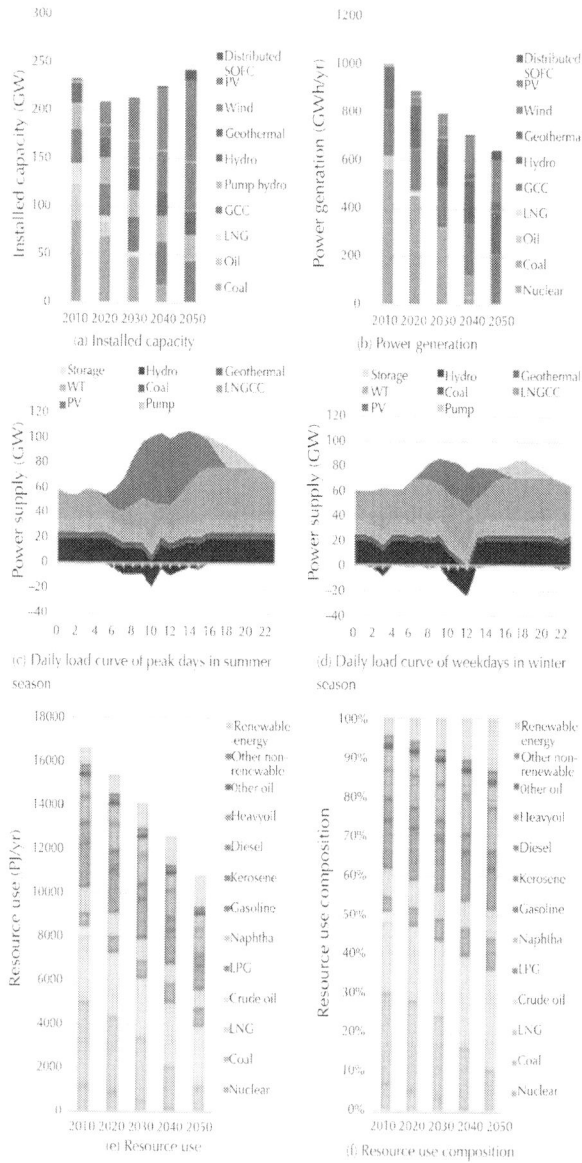

Figure 4: Composition of energy consumption in the power grid including centralized and distributed technologies (a–d), and in both supply- and demand-side systems (e and f) at the base technical scenario. (See also Fig. S5 as the electricity daily load curves in 2010 and 2050 estimated by EFM-FTO).

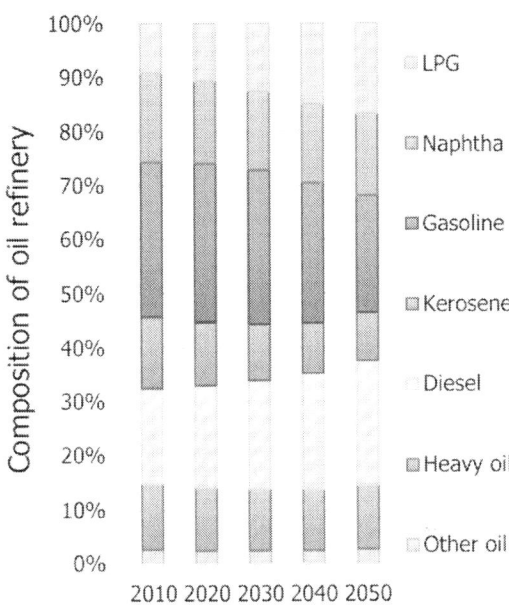

Figure 5: Composition of the demand for multi-products from oil refineries at the base technical scenario.

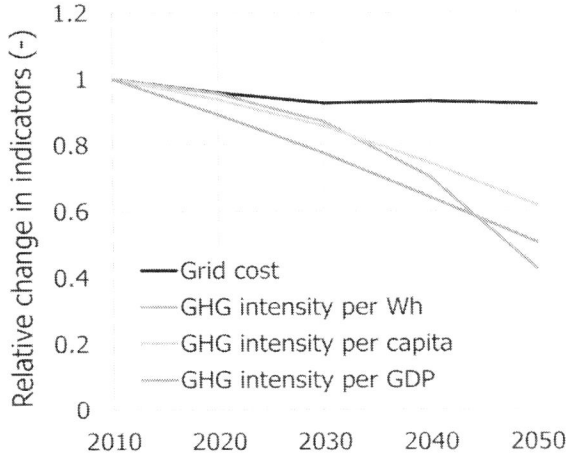

Figure 6: Relative change in grid cost and CO_2 intensities per capita and GDP at the base technical scenario.

Using Eq. (13) and $ActF_{yr}$ given in Table S2, GDP_{yr} was calculated, and the GHG intensity per GDP is shown in Fig. 6. It has a strong relationship with the GHG intensity per capita.

Technology Installation in the Residential Sector

Settings

Table 3 shows the societal and technical scenarios for the case study in the residential sector (R-Sce). The reference scenario was prepared to analyse the effects of R-Sce, including the change in population. R-Sce implements available technologies that are listed in Table 2. The technologies are divided into three categories: the replacement of centralized power generation (RepCentPow), the installation of energy-saving technology (EneSave) and the installation of SOFC (SOFC). R-Sce 2 to 6 consider the combination of technologies. The installed amount of PV and WT is decreased to half or increased to double the base technical scenario; i.e., 85 GW of PV and 50 GW of WT. R-Sce 9 adopts all available technology options. The effects of the order of technology implementation are analysed based on the GHG emission. Note that the detailed settings of technologies, such as the efficiency of SOFC, are the same as the base technical scenarios in the previous case study.

Table 3: Technical scenario settings for the residential sector in case study 2

		Reference scenario	R-Sce: scenarios in the residential sector								
			1	2	3	4	5	6	7	8	9
Societal parameter	Population change (Population)		✓	✓	✓	✓	✓	✓	✓	✓	✓
Technical parameter	Replacement of centralized power generation (RepCentPow)			✓		✓		✓	✓ Less PV and WT (Half)	✓ More PV and WT (Twice)	✓
	Installation of energy-saving technology (EneSave)				✓	✓			✓	✓	✓
	Installation of SOFC						✓	✓			✓
Result summary	GHG emission reduction ratio in residential sector in 2050 (%)	0	26	43	51	65	31	45	60	73	66
	Total GHG emission reduction ratio in 2050 (%)	0	5	15	9	19	5	16	16	26	20

Results

The result summaries of R-Sce are shown at the bottom of Table 3 and in Fig. 7, where the graphs are divided into three installation cases to compare R-Sce. Fig. 7(a) shows installation case 1. Comparing the reference and R-Sce1, the change in population has a significant effect on the GHG emission from the residential sector. Because the activity factor in the residential sector is population, the energy demand is directly affected by the change in population. A 26% reduction of GHG emission over 2010 can be achieved by the reduction in population. The replacement of centralized power generation also makes a large contribution to GHG emission reduction, which can reduce GHG emission another 17% over 2010. Rather than the replacement of centralized power generation, the installation of energy-saving technologies has a larger contribution to GHG emission reduction from the residential sector, as shown in Table 3. However, the replacement of centralized power generation makes a larger contribution to the reduction of total GHG emission, as shown in R-Sce2 in Table 3, because it can reduce the GHG emission from other sectors.

Fig. 7(b) shows installation case 2, where the installation of distributed co-generation and the replacement of centralized power generation are compared. The effect of installing SOFC has a relationship with substituted electricity from the power grid. As shown in R-Sce 5 and R-Sce 6 compared with R-Sce 1 in Fig. 7(b), the effect of SOFC installed under improved grid electricity, i.e., from R-Sce 2 to R-Sce 6, makes less contribution to GHG emission reduction than that from R-Sce 1 to R-Sce 5. This means that the order of installation of technologies may change their contribution to GHG emission reduction. Comparing R-Sce 9 with R-Sce 4 in installation case 1, the contribution of SOFC to GHG emission reduction is 1% reduction, which is much less than that in the comparison of R-Sce 5 with R-Sce 1. R-Sce 9 has the lowest GHG emission in installation case 2, if the additional renewable power sources are not acceptable. Installation case 3 analysing the effect of renewable power sources is shown in Fig. 7(c).

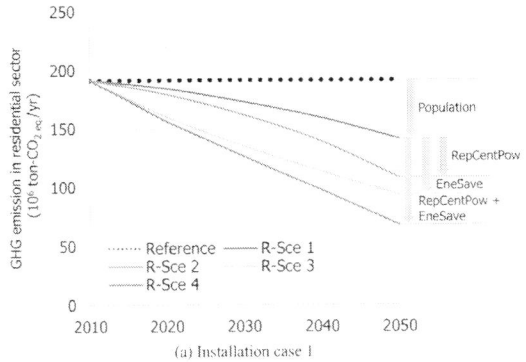

(a) Installation case 1

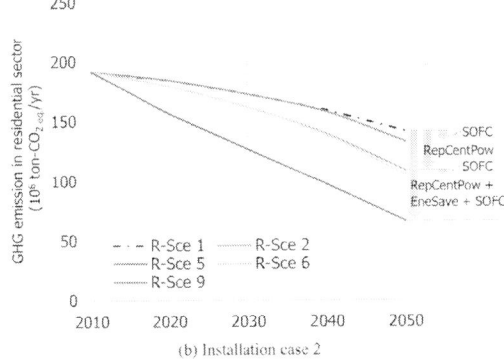

(b) Installation case 2

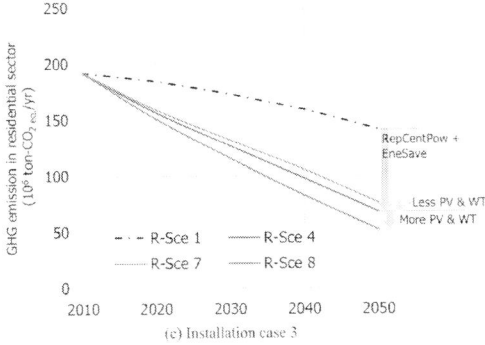

(c) Installation case 3

Figure 7: GHG emission in the residential sector for the technical scenarios shown in Table 3. The labels on the graphs represent the options adopted in each scenario; population: population change, RepCentPow: replacement of centralized power generation, EneSave: installation of energy-saving technology and SOFC: installation of SOFC.

The installation of renewable power generation technologies can contribute significantly to the reduction in GHG emission. In this regard, however, the amount of PV and WT in R-Sce 8, i.e., 170 GW of PV and 100 GW of WT, necessitates a much larger area for the installation. The additional reduction ratio, i.e., 7% from R-Sce 4 to R-Sce 8, should be carefully interpreted to indicate whether such aggressive installation of PV and WT is feasible.

Technology Installation in the Transportation Sector

Settings

The technical scenarios for the transportation sector (T-Sce) are shown in Table 4, which is divided into T-Sce for the comparison of HV with FCV (T-SceHF) and HV with EV (T-SceHE). In this case study, the penetration of new-generation vehicles was defined to reach a significantly high value; i.e., 90% of passenger or freight cars. This is because the objective of this case study is to demonstrate technology assessment by using the developed EFM-FTO. In T-SceHF, the technology options for hydrogen production were taken into account as scenarios. City gas and kerosene reforming and alkaline water electrolysis were the technologies adopted in demonstration facilities in Japan (e.g., [53] and [54]). The efficiency and required energy for producing hydrogen were extracted from existing reports (e.g., [53] and [54]). In the T-SceHF, the efficiency of the conversion of water to hydrogen is improved compared with T-SceHF with alkaline water electrolysis, from about 60% [53] to 85%. T-SceHFs 4 and 6 take into account the demand response to PV electricity supply. T-SceHFs 5 and 6 install twice as much PV as other T-SceHFs. In addition to the T-SceHFs, the T-SceHEs were generated for the comparison of EV with HV.

The penetration of HV and EV in these scenarios was defined as 90% of passenger cars. T-SceHE 2 takes into account the demand response to PV power supply. T-SceHE 3 considers one of the worst scenarios on electricity; i.e., the renewable power sources were not newly implemented after 2010. Total GHG emission was selected, and the break-even lines between FCV vs. HV and EV vs. HV were specified. Especially for FCV, the hydrogen production requires electricity, which can modify the electricity daily load curve significantly. The production of hydrogen can be applied for utilizing excess power supply from PV, which may result in the avoidance of much necessity of storages in power grid. The effects of such a kind of demand response were also analysed based on the installed or required capacities for meeting demands for power or hydrogen.

Table 4: Technical scenario settings for the transportation sector in case study 3

		T-SceHF						
		1	2	3	4	5	6	7
Penetration in passenger and freight cars (%)	FCV in FCV case	90 for all scenarios T-SceHF						
	HV in HV case	90 for all scenarios T-SceHF						
Hydrogen production		City gas reforming	Kerosene reforming	Alkaline water electrolysis (Improved conversion efficiency in T-SceHF 7 from about 60% to 85% in the electrolysis process)				
Demand response to PV		0	0	0	1	0	1	0

PV and WT installation in 2050 (PV, WT) (GW)		85, 50	85, 50	85, 50	85, 50	170, 100	170, 100	85, 50

Results

Figs. 8 and S6 show the obtained break-even lines for T-SceHFs and T-SceHEs respectively based on total GHG emission, the contour maps of which are shown in Fig. S7. An attainable region was given by the achievable fuel efficiency of new-generation vehicle technologies shown in Table S6. Based on the fuel efficiency, total GHG emission was calculated and compared for the scenarios of FCV vs. HV in all usage of vehicles, and EV vs. HV in passenger cars. The upper left direction of Fig. 8 means the optimal direction for FCV, while the opposite direction, i.e., lower right, is the direction for HEV. The break-even lines in Fig. 8represent the border-line of the priorities of FCV or HEV on GHG emission reduction. In Fig. 8, T-SceHFs 1 and 2 adopted reforming technologies, as shown in Table 4. Kerosene reforming requires more energy for reforming than city gas reforming. It results in a higher break-even line for T-SceHF2 than that of T-SceHF1, which means that the area where FCV has priority on GHG emission reduction in T-SceHF1 is larger than that in T-SceHF2. In this regard, however, the premises of selecting kerosene or city gas reforming for hydrogen production should be carefully taken into account. The kerosene can be supplied by existing infrastructures. It may result in a mix of technologies for producing hydrogen. At that time, the real break-even can lie between the lines of T-SceHFs 1 and 2 in Fig. 8. In addition to reforming technologies, alkaline water electrolysis was adopted for T-SceHFs 3 to 7. In the comparison of the break-even lines of electrolysis and reforming, the electrolysis line with base technologies on the power grid and conversion of water is located in the upper part of the attainable region. By improving the grid power efficiency, the break-even line moved to a lower region, as shown in T-SceHF 5 of Fig. 8. It was demonstrated that the demand response of hydrogen production to the PV power supply can slightly decrease GHG emission, and thus the break-even lines of T-SceHFs 4 and 6 were moved to a slightly lower region, as shown in Fig. 8. One of the effects of demand response was found by EFM-FTO and shown in Fig. 9, which shows the installed capacity of fossil-based thermal power plants and the required alkaline water electrolysis for the cases with/without demand response (see also Fig. S8 as the change in power demand by the demand response). Comparing them, the implementation of demand response can reduce the installed capacity of fossil-based thermal power plants, while it requires a larger capacity for hydrogen production. This is because the demand response of hydrogen production has become the receiver of large power supplied from PV. The net operation ratio of fossil-based fire power plants was improved at the expense of that of electrolysis. The border-line of T-SceHF7, where the electrolysis with the highly developed water conversion is assumed, demonstrated that the installation priority of electrolysis on GHG emission can be more than fuel conversion for hydrogen production. Similar results were demonstrated in the comparison of HEV with BEV. The power mix makes a significant difference in the break-even line between HEV and BEV (see also Fig. S6).

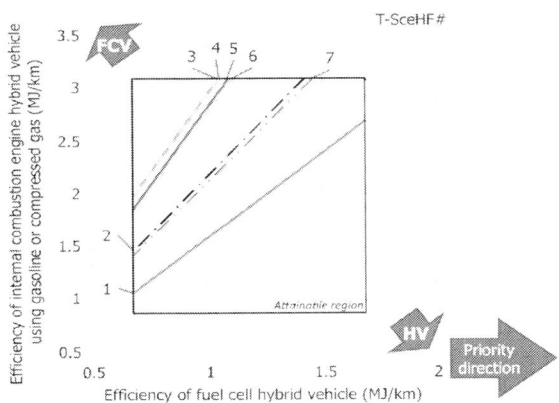

Figure 8: Break-even lines for technical scenarios on the FCV vs. HEV in transportation sector shown in Table 4.

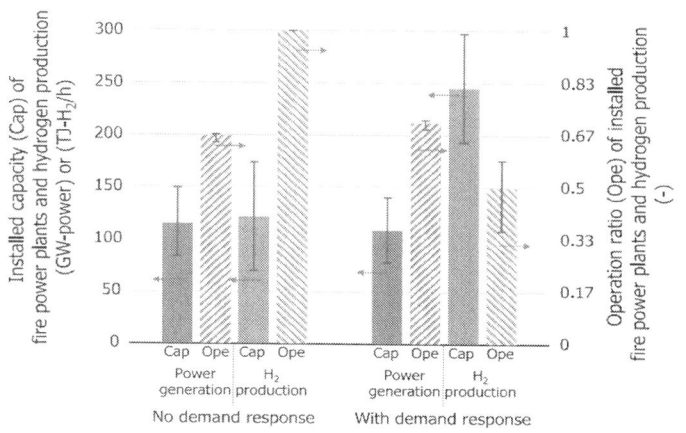

Figure 9: Change in the installed capacity (GW) of fossil-based fire power plant; coal, oil, LNG and LNGCC, and required hydrogen production capacity by alkaline water electrolysis (TJ/h) for T-SceF4.

DISCUSSION

As shown in the result of case study 1, EFM-FTO shows the performance and behaviour of energy systems on the capacity or energy flows of the installed technologies. GHG emission and the consumption of resources were the evaluated indicators of system performances as examples, which are basically so-called "less- or more-is-better" indicators. On the other hand, daily load curves or the installed capacity of technologies were adopted as the indices of the behaviour of the system, which does not always have a preferred direction. As discussed in case study 2, the installation order of technologies affects the net contribution of technologies to the reduction of GHG emission. As well as the installation order, the background conditions on the development and penetration of related technology made significant differences on technology assessments, as described in Fig. 8 of case study 3. Based on the outputted parameters of EFM-FTO, these visualizations can be conducted for all cases. Detailed technology functions can be taken into account by EFM-FTO. As shown in case study 3, the conversion efficiency of electrolysis will affect the break-even line significantly. Such deeper discussion on the technology development will be important for researchers and engineers of energy technologies to clearly set the research and development target [12]. To address this point, the functionals technology options should be broken down into parameters from the fundamental mechanisms of technologies for systems consisting of installed technologies. EFM-FTO has a simple structure based on the energy flows of 2010. The technology development in demand sectors are inserted into the functions of conversion efficiencies. The modularized technology model equipped in EFM-FTO facilitates the consideration of technology development. As well as technology development, societal changes were also taken into account. Societal changes here include not only the change in population but also the trend change in the selection of technologies; e.g., the electrification of air conditioning. The penetration of technologies can also be affected by the changes of lifestyle or the sense of values.

EFM-FTO can be regarded as a technology-oriented engineering model. Energy system models generally can be divided into three types: an engineering model mostly including optimization mechanisms, a computable general equilibrium (CGE) model, and a simulation-oriented model [19]. EFM-FTO requires a set of parameters on technology development and penetration. Such a set of parameters should be able to address the sophisticated and up-to-date understanding of a wide range of technologies, which can be collected from experts and stakeholders of the energy systems. The data set in this paper is extracted from interviews and listening to experts, including researchers and engineers on energy technologies in Japan, and through workshops for editing a book on energy technology roadmapping [12]. EFM-FTO has the potential to be employed as a communication tool for gathering up-to-date information on feasible technologies. The model was developed with a modularized structure enabling the decomposition or aggregation of the boundaries of parameterized technologies. This type of modelling approach has been applied in the systematic chemical process design [55] and described as ontology-based structures [56]. It addresses detailed engineering on energy technologies such as the installation of energy-saving and co-generation technologies in industries. EFM-FTO can describe the technical aspects of energy scenarios, which have not been supported in detail by the usual economy system-based energy models. The accumulation of the functionals of technology options in energy systems was established by EFM-FTO and can show the attainable regions associated with the installation of technology options by giving indicators as shown in case studies.

The installation of a distributed energy system has become an issue associated with a more sustainable and resilient society especially in Japan after 3.11, 2011 (e.g., [1]). Fig. 10 shows a comparison of the share of distributed and centralized power generation outputted from EFM-FTO on the scenario in the base scenario in Section 3.2.1. Fig. 10 (a) shows the power supplied from distributed and centralized power generation systems. The distributed system includes the SOFC installed in residential areas, PV power generation and power systems installed in industries. The ratio of these categories of power sources is shown in Fig. 10 (b). The decentralized systems have about 36% of the share of power supply in 2050. It is more than four times the share of distributed power in 2010. The increase of decentralized power sources can decrease the energy loss associated with the conversion of chemical energy of fuels [57]. Collaboration of multi-industrial factories, the so-called industrial symbiosis [58], can achieve a significant reduction in total fuel consumption due to the advanced material and heat integration management. In addition, the distribution of power sources may have a strong relationship to the vulnerability of the energy system [59] towards disasters destroying energy infrastructures. The parameterization and quantification of the behaviour and performance of the energy system by EFM-FTO may be a support for discussing expected risks on the future energy system.

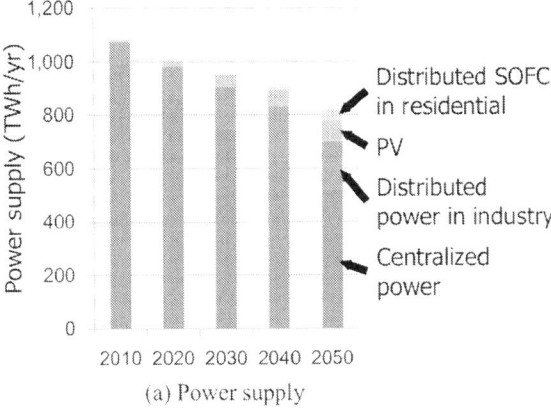

(a) Power supply

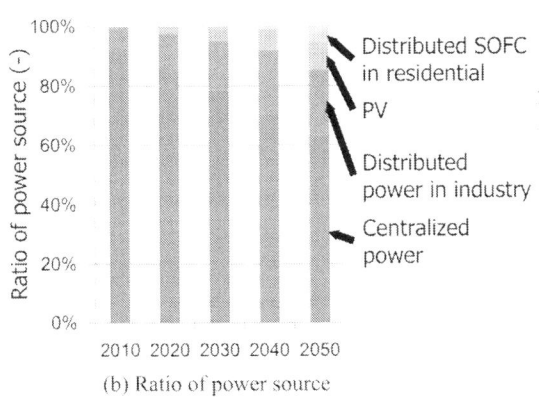

(b) Ratio of power source

Figure 10: Comparison of the share of the distributed power generation system with centralized power generation.

The energy intensities on societal parameters were specified by the future energy flows. Eq. (14) schematically represents the relationship among key parameters of energy intensity. Energy demand per activity factor shown was assumed to be constant in the case studies in this paper. As shown in Eq. (12), GDP per activity factor was also assumed to be constant. Based on this assumption, Eq. (14) was derived. The left-hand side describes the energy intensity on GDP. It is expanded by energy demand and the activity factor. FuelC/EneD was defined as the reciprocal of Eff, the conversion efficiency from fuel to energy. EneD/ActF was a utilized assumption in the analysis of case studies. Hence, the energy intensity is dependent on the changes in Eff, which is shown in Eq. (15). An example of the change in energy intensity is shown in Fig. 6. The technology installation produces changes in Eff resulting in a change of energy intensity.

$$\frac{FuelC_{yr}}{GDP_{yr}} = \frac{FuelC_{yr}}{EneD_{yr}} \cdot \frac{EneD_{yr}}{ActF_{yr}} \cdot \frac{ActF_{yr}}{GDP_{yr}} \qquad (14)$$

$$\frac{FuelC_{yr}}{GDP_{yr}} = \frac{1}{Eff_{yr}} \cdot \frac{EneD_{yr}}{ActF_{yr}} \cdot \frac{ActF_{yr}}{GDP_{yr}}. \qquad (15)$$

The EFM-FTO has limitations on the scenario analysis because of its characteristics. It is an energy flow model, which is represented as a static flow model, not directly considering the construction/decommission of infrastructures. While construction/manufacturing industries are also included in the industrial sector in EFM-FTO, materials production, assembling, system construction and decommission are not considered explicitly and supply chains are not considered in the present version of the model. The amount of construction or other product manufacturing can be calculated inversely on the basis of the assumption that the same energy consumption leads to the same amount of production; i.e., kg-product/J-energy. In this regard, such energy efficiency for product manufacturing can be improved by technology implementation. In the analysis of this paper, the activity factor kept increasing during 2010 to 2050, which means that the domestic production amount also increases. While the change of industrial structure, e.g., from heavy industries to service industries, may significantly decrease the energy consumption per GDP or other intensities, the activities of industrial and commercial sectors in this paper were increased for comparing scenarios without excess expectation of GHG emission reduction by structural changes in industries. If there is a shortage or excess supply of products, it was assumed in EFM-FTO that the products are imported or exported, respectively. To fulfil a wider and more global energy flow analysis, much more statistical data should be accessible on the multi-country trades composing the supply chain.

EFM-FTO can be employed as a tool for scenario analysis of energy flow from various viewpoints. For example, economic aspects of energy system have been discussed for the renewable options utilized in power grid [60] or for fuels [61]. In this paper, only power-grid cost was calculated for specifying the operational conditions of power plants as shown in Section 2.2, while EFM-FTO outputs required information for calculating the total costs due to the installation of new power sources such as PV or distributed FC. The discussion of the cost to counter the frequency fluctuation of the power grid caused by intermittent power sources such as PV will be a remaining target. While the cost must be covered by some entities, it may not be always the case to include such additional costs into the power grid cost. As the further researches by applying the developed EFM-FTO, systems design assuming the stake-holders potentially involved in the payment of such costs is one possiblity. Other important issues in energy system such as energy security (e.g., [62]) also requires the consideration of multiple aspects (e.g., [63]) and involvement of a variety of stake-holders. The developed EFM-FTO should be to provide some of quantitative bases required for the discussion among stake-holders.

Developed EFM-FTO has a modularized structure represented as functionals of technology options. The geographical scope can be modified by using multi-country energy demand records. At the same time, the scope of EFM-FTO can be narrowed by using region-wide energy demand records. Inside of a region or country, e.g., Japan, the segments of power grid can also be considered by EFM-FTO. As discussed by Koyama et al., the transmission among different sub-region inside of Japan has become issues for sharing and stabilizing power supply [22]. In the further discussion based on developed EFM-FTO, the construction of power substation can be taken into account by dividing the power supply model dispatching power sources. Note that the demand side model has region-specific energy demand. In addition to such geographical scope, the depth of the functions of technology options can also be changeable by preparing technical information and scenarios. For example, the smart city or other related projects (e.g., [64]) have potentials to reduce the fossil fuel consumption by smart use of the energy. Another important potential is to mitigate the frequency fluctuation of the grid caused by PV and WT, which will be a support for the large-scale introduction of those intermittent power sources. These viewpoints can be taken into account in future study by extending modules to EFM-FTO.

CONCLUSIONS

We have developed a model of energy flow implementing technology options as functionals of technology options. Based on the model, the development and replacement of technologies installed in the current energy system in Japan were quantitatively analysed by setting scenarios as sets of parameters associated with societal and technical changes. The demand-side module of the energy system has submodules representing demand sectors; i.e., residential, commercial, industrial and transportation. Each submodule was modularized into component energy demands such as heating, cooling or transferring, which were fulfilled by fuel and appliances. With the understanding of the current energy demand, future primary and secondary energy demands were estimated by modelling available technology options. The supply side of the energy system, on the other hand, is designed to address the hydrogen and electricity as secondary energy demand calculated at the demand side submodule. For hydrogen production, the fluctuation of production inventories was quantitatively analysed. The selection of technology options for secondary energy affected the priorities of demand-side technologies such as vehicle technologies. Centralized power generation also has large impacts on the selection of demand side technologies. The need for a distributed co-generation system such as SOFC in the residential sector or electrification of heating appliances was obvious.

Scenario analysis applying the developed model showed multi-aspects of the energy system, such as performances or behaviours. Case studies demonstrated the scenario analysis based on the developed model. As the performances, items such as GHG emission or fossil resource consumption were shown for demand sectors or resource types. Electricity daily load curves for representative days in 2010 to 2050 were analysed, and power mixes were specified considering installable capacities of power generation plants based on load-following capacities as the behaviours of the energy system. The priorities of available technologies considering the conditions of other related systems were analysed by the developed model as shown in the case study on vehicle technologies. The generation of secondary energies such as hydrogen and electricity has a strong connection with the future priorities of vehicle technologies, the fuel efficiency of which also has a possible range because of the promotion of development. Quantitative analysis enabled by the developed model facilitates a detailed discussion of technical scenarios.

Towards the involvement of researchers and engineers in the discussion and communication of multi-aspects of the energy system, the parameters that they can decide or affect by their efforts should be inter- or intraconnected with the results of scenario analysis. Fundamental researchers and engineers are involved through the discussion and specification of the physical and chemical mechanisms of devices; e.g., the efficiency of SOFC or the conversion of primary into secondary energies. The effects of district heating [20] and industrial symbiosis [57] are interconnected with the modules for the residential and industrial sectors, respectively. Macroscopic changes, on the other hand, such as the changes in industrial structure and supply chains, should also be able to be addressed. Analytical methods for such supply chain or industrial structure [59] should also be intraconnected with technology-based energy models.

The developed energy model can receive scenario parameters from various viewpoints, including fundamental chemists, industrial engineers, local governments and policymakers by converting their scenarios into parameters. All analyses by the model are based on the performance of the current energy system. This means that forecasting by the model is enabled as a set of forecasts of microenergy demand/supply in the current system. By excluding the energy demand change directly affected by the economic situation, the potential of technical scenarios can be analysed simply without the social elasticity of energy demand. In this regard, however, modifying the activity factors for subdivisions of demand sectors enables the incorporation of economic scenarios.

ACKNOWLEDGEMENTS

The authors are grateful to Dr. Keiko Fujioka, Profs. Takao Nakagaki, Yukitaka Kato, Masahiko Matsukata, Mitsuhiro Kubota, Yuya Kajikawa, Yasuhiro Fukushima and all authors of Energy beyond '20 – Futures Built by Feasible Technologies for their fruitful discussions and support. The Committee on Future Energy and Social Systems, Center for Strategic Planning, the Society of Chemical Engineers, and Japan is also gratefully acknowledged for the discussion of future energy systems. Activities of Inamori Frontier Research Center are supported by KYOCERA Corporation. Activities of the Presidential Endowed Chair for "Platinum Society" at the University of Tokyo are supported by the KAITEKI Institute Incorporated, Nippon Telegraph and Telephone Corporation, Fujifilm Holdings Corporation, Mitsui Fudosan Corporation and LIXIL Corporation.

REFERENCES

1.   Fukushima Y, Kikuchi Y, Kajikawa Y, Kubota M, Nakagaki T, Matsukata M, et al. tackling power outages in Japan: the earthquake compels a swift transformation of the power supply. J Chem Eng Jpn 2011; 44:365–9.

2.   McDowall W, Eames M. Forecasts, scenarios, visions, backcasts and roadmaps to the hydrogen economy: a review

of the hydrogen futures literature. Energy Policy 2006; 34:1236–50.

3.   Chen IC, Fukushima Y, Kikuchi Y, Hirao M. A graphical representation for consequential life cycle assessment of future technologies. Part 1: Methodological framework. Int J Life Cycle Assess 2012; 17:119–25.

4.   Chen IC, Fukushima Y, Kikuchi Y, Hirao M. A graphical representation for consequential life cycle assessment of future technologies. Part 2: Two case studies on choice of technologies and evaluation of technology improvements. Int J Life Cycle Assess 2012; 17:270–6.

5.   Fukushima Y, Shimada M, Kraines S, Hirao M, Koyama M. Scenarios of solid oxide fuel cell introduction into society. J Power Sources 2004; 131:327–39.

6.   Kannan R, Strachan N. Modelling the UK residential energy sector under longterm decarbonisation scenarios: comparison between energy systems and sectoral modelling approaches. Appl Energy 2009; 86:416–28.

7.   Tonini D, Astrup T. LCA of biomass-based energy systems: a case study for Denmark. Appl Energy 2012; 99:234–46.

8.   Gebremedhin A. Introducing district heating in a Norwegian town – potential for reduced local and global emissions. Appl Energy 2012; 95:300–4.

9.   Takeshita T. A strategy for introducing modern bioenergy into developing Asia to avoid dangerous climate change. Appl Energy 2009; 86:S222–32.

10. Kajikawa Y, Kikuchi Y, Fukushima Y, Koyama M. Utilizing risk analysis and scenario planning for technology roadmapping. In. Cetindamar D, Daim T, Beyhan B, Basoglu N, editors. Strategic planning decisions in the high tech industry. Heidelberg: Springer; 2013. p. 231–244.

11. Ministry of Economy, Trade and Industry (METI), Japan. Strategic Technology Roadmaps 2010 (Gijutu Senryaku Mappu 2010); 2010. <http:// www.meti.go.jp/policy/

economy/gijutsu_kakushin/kenkyu_kaihatu/   str2010.html>
[accessed 25.03.14].

12.   Kato Y, Yasunaga H, Kashiwagi T, editors. Energy beyond '20
– Futures Built by Feasible Technologies (Jissokano na Energy
Gijyutu de Kizuku Mirai – Honebuto no Energy Road Map
2 – (original title in Japanese)). KagakuKogyo-sha: Kawasaki;
2010.

13.   Kajikawa Y, Yoshikawa J, Takeda Y, Matsushima K. Tracking
emerging technologies in energy research: toward a roadmap
for sustainable energy. Technol Forecast Soc 2008; 75:771–
82.

14.   COMMEND (Community for Energy, Environment and
Development). An Introduction to LEAP; 2012. <http://www.
energycommunity.org/documents/ LEAPIntro.pdf> [accessed
25.08.12].

15.   National Institute for Environmental Studies, Japan (NIES);
2012.   <http:// www-iam.nies.go.jp/aim/infomation.htm>
[accessed 25.08.12].

16.   Loulou R, Goldstein G, Noble K. Documentation for the
MARKAL family of models; 2004. <http://www.etsap.org>.

17.   Ichinohe M, Endo E. Analysis of the vehicle mix in the
passenger-car sector in Japan for CO_2 emissions reduction by
a MARKAL model. Appl Energy 2006; 83:1047–61.

18.   Lee DH, Park SY, Hong JC, Choi SJ, Kim JW. Analysis of the
energy and environmental effects of green car deployment by
an integrating energy system model with a forecasting model.
Appl Energy 2013; 103:306–16.

19.   Agency). A report on energy model for constructing an
abundant and sustainable society. (Yutakana Jizokusei Syakai
Kouchiku notameno energy model Houkokusyo.) CRDS-
FY2011-WR-11 2011; 2011.

20.   Wetterlund E, Söderström M. Biomass gasification in district
heating systems— the effect of economic energy policies.
Appl Energy 2010; 87:2914–22.

21.   Nishimura K, Hondo H, Uchiyama Y. Comparative analysis of

embodied liabilities using an inter-industrial process model: gasoline- vs. electropowered vehicles. Appl Energy 2001; 69:307–20.

22. Koyama M, Kimura S, Kikuchi Y, Nakagaki T, Itaoka K. Present status and points of discussion for future energy systems in Japan from the aspects of technology options. J Chem Eng Jpn 2014. http://dx.doi.org/10.1252/ jcej.13we345.

23. Agency for Natural Resources and Energy (ANRE), METI, Japan. Table of the FY2010 comprehensive energy statistics; 2012.

24. Greenhouse Gas Inventory Office of Japan. The GHGs Emissions Data of Japan; 2013. <http://www-gio.nies.go.jp/aboutghg/nir/nir-e.html> [accessed 25.03.14].

25. Kannan R. The development and application of a temporal MARKAL energy system model using flexible time slicing. Appl Energy 2011; 88:2261–72.

26. Chubu Electric Power Company (Chubu-EPCO). Past records of power demand; 2013. <https://www.chuden.co.jp/> [accessed in 2013].

27. Chugoku Electric Power Company (Chugoku-EPCO). Past records of power demand; 2013. <https://www.chuden.co.jp/> [accessed in 2013].

28. Hokkaido Electric Power Company (HEPCO). Past records of power demand; 2013. <http://www.hepco.co.jp/> [accessed in 2013].

29. Hokuriku Electric Power Company (Hokuriku-EPCO). Past records of power demand; 2013. <http://www.rikuden.co.jp/> [accessed in 2013].

30. Kansai Electric Power Company (KEPCO). Past records of power demand; 2013. <http://www.kepco.co.jp/> [accessed in 2013].

31. Kyushu Electric Power Company (Kyushu-EPCO). Past records of power demand; 2013. <http://www.kyuden.co.jp/> [accessed in 2013].

32. Shikoku Electric Power Company (Shikoku-EPCO). Past records of power demand; 2013. <http://www.yonden.co.jp/> [accessed in 2013].

33. Tohoku Electric Power Company (Tohoku-EPCO). Past records of power demand; 2013. <http://www.tohoku-epco.co.jp/> [accessed in 2013].

34. Tokyo Electric Power Company (TEPCO). Past records of power demand; 2013. <http://www.tepco.co.jp/index-j.html> [accessed in 2013].

35. Agency for Natural Resources and Energy (ANRE), METI, Japan; Estimation of the peak electricity demand structure in summer season. (Kakisaidaidenryokushiyoubi no Juyokouzousuikei); 2014. <http:// www.meti.go.jp/setsuden/20110513taisaku/16.pdf> [accessed in 2014].

36. The Federation of Electric Power Companies of Japan (FEPC). Electricity Demand Record. (Denryoku Juyo Jisseki); 2014. <http://www.fepc.or.jp/   library/data/demand/index.html> [accessed in 2014].

37. Thermal and Nuclear Power Engineering Society. Handbook of Thermal and Nuclear Power Plants (Karyoku genshiryoku hatudensyo setsubi youran); 2008.

38. Ministry of Finance Japan (MOFJ). Trade Statistics of Japan; 2013. <http:// www.customs.go.jp/toukei/info/index_e.htm> [accessed 12.12.13].

39. International Energy Agency (IEA). 2012 Key World Energy Statistics", International Energy Agency, Paris, France (2012). Energy technology   perspectives 2010. International Energy Agency, Paris, France; 2010.

40. Cost Verification Committee (CVC). Cabinet Office, Government of Japan. Report of the cost verification committee (Kosuto tou kensho iinkai repohto); 2011.

41. Kato Y, Matsukata M, Nakagaki T, Kikuchi Y, Kajikawa Y, Koyama M, Kubota M. Development of future energy system in Japan by feasible technologies (Jisso   Kano na Gijutu

niyoru Wagakunino Mirai Energy System no Kouchiku.) Final reports of environment research and technology development fund by   ministry of the environment, Japan; 2012.

42. Koyama M, Kraines S, Tanaka K, Wallace D, Yamada K, Komiyama H. Integrated model framework for the evaluation of an SOFC/GT system as a centralized power source. Int J Energy Res 2004; 28:13–30.

43. IPSS (National Institute of Population and Social Security Research, Japan). Population Projections for Japan; 2012.

44. Chua KJ, Chou SK, Yang WM. Advances in heat pump systems: a review. Appl Energy 2012; 87:3611–24.

45. Kikuchi Y, Hirao M, Sugiyama H, Papadokonstantakis S, Hungerbühler K,   Ookubo T, et al. Design of recycling system for poly (methyl methacrylate)   (PMMA). Part 2: Process hazards and material flow analysis. Int J Life Cycle   Assess 2014; 19:307–19.

46. Ueno T, Kitahara H. Development of heat characteristic model of home user air   conditioner – Part 3: Characteristics model for heating period. CRIEPI report   R10009 (2011). Development of heat characteristic model of home user air   conditioner – Part 3: Characteristics model for heating containing defrost   period. CRIEPI report R11017; 2012.

47. Advanced Cogeneration and Energy Utilization Center Japan; 2013. <http:// www.ace.or.jp/> [accessed 28.11.13].

48. Japan Center for Climate Change Actions. A fact sheet of home electric   appliances for energy-saving (Syo enerugi kaden fakuto shihto), 2007. A fact   sheet of residences for energy-saving (Syo enerugi jutaku fakuto shihto); 2006.

49. Ichimoto K, Nakamura H, Kawai R. "More convenience" pursuit of large capacity and energy-saving for refrigerators. Hitachi Rev 2010; 92:762–3.

50. Ikeboh Y. The latest technology for refrigerator. J Inst Elect Eng Jpn 2012; 132:539–41.

51. Japan Automobile Dealers Association. Statistics of vehicle sales; 2013. <http:// www.jada.or.jp/> [accessed in 2013].

52. Kimura S. A methodology study for assessment of social impact by introducing next generation vehicles, Ph.D. thesis. Kyushu University; 2013. p. 92–112.

53. JARI (Japan Automatic Research Institute). A report of the analysis on overall efficiency and GHG emissions; 2011.

54. JARI (Japan Automatic Research Institute). The result of hydrogen station.   Hydrogen Station Ryoiki Shiken kekka; 2014. <http://www.jari.or.jp/Portals/0/   jhfc/data/seminor/ fy2004/pdf/4_H16JHFC.pdf> [accessed March 2014].

55. Klatt KU, Marquardt W. Perspectives for process systems engineering—   personal views from academia and industry. Comput Chem Eng   2009; 33:536–50.

56. Marquardt W, Morbach J, Wiesner A, Yang A. OntoCAPE: a re-usable ontology for chemical process engineering. Heidelberg: Springer; 2010.

57. Bouffard F, Kirschen D. Centralised and distributed electricity systems. Energy Policy 2008; 36:4504–8.

58. Chertow M, Ehrenfeld J. Organizing self-organizing systems: toward a theory of industrial symbiosis. J Ind Ecol 2011; 16:13–27.

59. McLellan B, Zhang Q, Farzaneh H, Utama NA, Ishihara KN. Resilience, Sustainability and risk management: a focus. Energy 2012; 3:153–82.

60. Koo J, Park K, Shin D, Yoon ES. Economic evaluation of renewable energy   systems under varying scenarios and its implications to Korea's renewable   energy plan. Appl Energy 2011; 8:2254–60.

61. Heyne S, Harvey S. Assessment of the energy and economic performance of   second generation biofuel production processes using energy market   scenarios. Appl Energy 2013; 101:203–12.

62. Gracceva F, Zeniewski P. A systemic approach to assessing

energy security in a low-carbon EU energy system. Appl Energy 2014; 123:335–48.

63. Kiriyama E, Kajikawa Y. A multilayered analysis of energy security research and the energy supply process. Appl Energy 2014; 123:415–23.

64. Yamagata Y, Seya H. Simulating a future smart city: an integrated land use energy model. Appl Energy 2013; 112:1466–74.

A Comparative Study of Ammonia Energy Systems as a Future Energy Carrier, With Particular Reference to Vehicle Use in Japan

Daisuke Miura and Tetsuo Tezuka

The Department of Socio-Environmental Energy Science, Graduate School of Energy Science, Kyoto University, Yoshida Hon-machi, Sakyo-ku, Kyoto 606- 8501, Japan

ABSTRACT

The choice of secondary energy carriers, such as electricity, hydrogen and ammonia, influences not only economic and environmental performances but also the reliability of an entire energy system. This article focuses on ammonia because of its excellent property in energy storage, and assesses the relative advantages of several ammonia energy systems for vehicle use in Japan by estimating energy efficiency, CO_2 emissions, and the supply cost of several ammonia energy paths, which are then compared with alternative paths using different energy carriers including hydrogen and electricity. The article also discusses inherent merits and challenges of ammonia energy systems and identifies directions for future research and development. Using ammonia as an energy carrier was demonstrated to be competitive in terms of efficiency, CO_2 emissions and supply cost for energy systems requiring fairly large numbers of storage days. This assessment shows that the use of ammonia in an energy system can improve the continuity of the energy supply in a country or region with insecurity of supply. On the other hand, we argue that further technical improvements and cost reduction associated with both conventional and unconventional ammonia production is imperative for using ammonia in a normal energy system.

INTRODUCTION

The energy supply crisis which was caused by the great east Japan earthquake has raised a necessity for Japan to rebuild a robust energy supply system in the long term [1] and [2]. The new energy system should not only pursue economic competitiveness and environmental harmonization, but also stress on the contribution to a reliable energy supply, which enhances resilience to external supply shocks.

Choice of secondary energy carrier such as electricity, hydrogen, and ammonia influences economic and environmental performances as well as the reliability of total energy supply systems. Electricity, one of the most common energy carriers, is usually considered to be both economically and environmentally competitive when utilized with the benefit of smart grid concept and zero carbon feature of renewable energy sources. But storage is a significant drawback because it is usually costly [3]. Supply and demand must always be matched simultaneously, and a slight imbalance could cause a widespread blackout. In addition, the current widespread suspension of nuclear power plants in Japan has resulted in increased use of fossil fuel power plants, leading to deterioration of environmental competitiveness in terms of CO_2 emissions. Hydrogen as an energy carrier exerts competitive advantage when large-scale energy storage is required in an energy system because hydrogen can be more conveniently stored than electricity. However, several issues including economics and technological development particularly in the area of fuel cells need to be addressed for hydrogen to become commercially viable [4].

This article pays a special attention to ammonia as a promising energy carrier. The technologies associated with the production, transportation and storage of ammonia are well established, resulting in a convenient and low-cost energy supply chain. The potential advantages of large-scale storage of ammonia could be harnessed to maintain energy delivery to final consumers without causing interruption even in the case of supply disruption such as the great east Japan earthquake.

Potential use of ammonia as an energy carrier has been studied mostly in engineering fields including production, storage and utilization. Apart from the efforts of improving conventional Harbor-Bosch process[5], decentralized ammonia production technology with less CO_2 emissions is being developed. For example, Ito et al. proposed electrochemical ammonia synthesis from water vapor and nitrogen [6]. Nishibayashi et al. exhibited the catalytic reduction of nitrogen to ammonia in the presence of a molybdenum complex under atmospheric pressure [7]. However, both methods are still in research stage, thus requiring further clarification for fundamental reaction mechanism before moving toward future commercialization. As to the storage technology, although the storage of ammonia in steel vessel is already the well commercialized method, a new attention is paid to the ammonia-absorbing materials such as $Ca(NH_3)_8Cl_2$ and $Mg(NH_3)_6Cl$, which could mitigate smell and toxic issues specific to ammonia[8] and [9]. For the utilization (end-use), research efforts have been focused on the following two technologies. First, ammonia internal combustion engine has been researched aiming to vehicle use. Ammonia is thought to be a clean fuel since its direct combustion (i.e., exoergic reaction) creates only water and nitrogen. Research is focusing on improving ammonia's ignition characteristics by identifying optimal mechanical engine systems and studying effective combustion improver such as hydrogen[10] and [11]. Second, ammonia can be used as a feedstock to fuel cell for electricity generation, by directly supplying ammonia to solid oxide fuel cell (SOFC) or by reforming ammonia to produce hydrogen to be supplied to polymer electrolyte fuel cell (PEFC) [12], [13] and [14]. Technical challenges remain in each ammonia utilization technology. In PEFC, existence of tiny amount of unreformed ammonia causes polymer electrolyte membrane to rapidly deteriorate. In SOFC, in spite of the advantage of the lower working temperature at 673–873 K, generation performance is still inferior to conventional SOFC, which works at higher temperature using conventional fuel such as natural gas. This implies further technical improvement including searching for catalyst is required.

On the other hand, studies on economic and environmental aspects of ammonia energy systems have been limited to the ones dealing with qualitative assessment, or cost estimation of a fraction of ammonia supply chain in the United States [15], [16], [17] and [18]. More importantly, as far as the authors are concerned, few studies have been carried out on comprehensive evaluation framework of whole ammonia energy supply chain taking production, transport, storage and utilization into account, which makes us difficult to assess the general competitiveness of ammonia-based energy systems against electricity and/or hydrogen based energy systems. In addition, the lack of such the evaluations raises difficulty in discussing the requirement that various ammonia related technologies must achieve for future realization of the ammonia energy systems.

Based on this background, this article first proposes a simple methodology of economic and environmental analysis framework for ammonia-based energy supply systems that take the whole ammonia supply chain into account. Using the methodologies, this article aims to assess the comparative advantage of several ammonia systems for vehicle use in Japan by estimating energy efficiency, CO_2 emissions and supply cost of several ammonia energy paths, and then comparing them with alternative paths using different energy carriers including hydrogen and electricity. The article also discusses the inherent merits and challenges of ammonia energy systems, and identifies the directions for future research and development (R&D).

This article is structured as follows. In Section 2, a standard methodology is developed to remove the biases associated with both the data collection and the calculation assumptions made in comparing different energy supply pathways using various types of energy carriers including ammonia. Section 3presents results on the comparative advantage of ammonia energy systems against those with the other carriers using three types of indicator. It also discusses R&D issues which need to be addressed so that energy supply systems using ammonia can exhibit clearer merits in terms of energy efficiency, economics and environmental aspects in the future. Conclusions and future works are summarized in Section 4.

METHODOLOGIES

Energy Supply Pathway

The ammonia economy will first emerge in the vicinity of ammonia production plants with surplus capacity. In Japan, the ammonia production capacity was 2.06 million tons in 2004, while actual production has consistently decreased to 1.17 million tons in 2010 due to the reduced domestic demand [19]. This means the redundant capacity of 0.89 million tons (which is equivalent to approximately 1.65×10^{10} MJ in terms of the lower heating value), can be used to produce the ammonia energy career. If fully utilized, this quantity is equivalent to annual fuel requirement for approximately one million ammonia internal combustion vehicles[16]. The first introduction of ammonia supply infrastructure will appear in the vicinity of ammonia production plants, considering that the shorter the distance for transporting ammonia is between production premise and consumption premise, the lower the total supply cost is. Meanwhile, it can be also said that the supply of this type of ammonia can mobilize only approximately 0.7% of the total passenger transport demand in Japan as of 2010 [20]. Therefore, in the long term, the aforementioned new decentralized synthesis methods will need to be developed and commercialized toward a more widespread and more environmentally friendly pathway to supply ammonia.

Table 1 shows the energy supply pathways that were examined in this study. For each pathway, the technologies associated with production, transportation, storage and utilization of energy carrier have been selected by taking into account the current status of technological developments. In general, energy supply path is categorized as either "off-site" or "on-site". Off-site pathway assumes that energy carrier is economically produced in a large-scale plant, but cost for transportation is incurred because the production location is usually far from the demand location. On-site pathway does not require transportation since the location of production coincides with that of usage, but the scale of production plant is reduced, which incurs a higher production cost. Both the off-site and on-site pathways require storage facilities in order to compensate for the daily and seasonal deviations between supply and demand.

Table 1: Types of energy supply paths

Type	Conversion	Transportation	Storage	Utilization
Off-site system				
AES1	Haber–Bosch	HP[a] truck	HP storage tank	Ammonia internal combustion vehicle
HES1	Byproduct from refinery	HP truck	HP storage tank	Fuel cell vehicle

On-site system				
AES2	Electrochemical synthesis	N/A[b]	HP storage tank	Ammonia internal combustion vehicle
HES2	Methane reforming	N/A	HP storage tank	Fuel cell vehicle
HES3	Electrolysis	N/A	Rapid charger	Electric vehicle
Electricity system				
EES	Grid electricity		Rapid charger	Electric vehicle

[a]HP: high pressure.

[b]N/A: not applicable.

Ammonia energy system 1 (AES1) represented a commonly adopted off-site pathway where ammonia is produced from natural gas in a large-scale production plant using a well-known Haber–Bosch process [21]. The produced ammonia is transported by a truck with pressure vessel and stored in a cylindrical pressure tank. On-site ammonia energy path has not been commercialized yet, but several novel production techniques have been proposed including electrochemical ammonia synthesis from water vapor and nitrogen, and catalytic reduction of nitrogen to ammonia in the presence of a molybdenum complex, as stated in Section 1[6] and [7]. In this study, the electrochemical method was assumed to become technically viable in the near future, which was nominated as the ammonia energy system 2 (AES2).

Merits of the aforementioned AESs were examined through the comparison with energy supply pathways based on either hydrogen or electricity. An off-site hydrogen energy system (HES1) assumed the usage of hydrogen as a byproduct in refinery. As on-site hydrogen systems, we considered two forms: HES2 provided hydrogen reformed from natural gas to fuel cell vehicle (FCV) at a hydrogen fueling station, and HES3 assumed hydrogen to be produced by water electrolysis at the aforementioned fueling station. As the electricity-based energy system (EES), we assumed only one off-site pathway in which grid electricity is simply supplied to an electric vehicle (EV) by an electric charger. On-site electricity systems that solely rely on distributed generations without utilizing existing grid electricity are considered to be economically unfeasible, given the fully established grid distribution network in Japan.

The Evaluation Indicators

The advantages of AES were compared using three indicators: the primary energy consumption per energy carrier supplied (MJ/MJ), the CO_2 emissions per energy carrier supplied (kg-CO_2/MJ), and the supply cost per energy carrier supplied (JPY/MJ). As expressed in Fig. 1, the primary energy consumption is defined as the sum of primary energy used for the primary fuel and the process of conversion, transportation and storage. Fig. 1 shows that the primary energy consumption per energy carrier supplied is calculated as $(X_1 + A_1 + A_2 + A_3)/X_4$. The CO_2 emissions are the amount of CO_2 associated with the consumption of the aforementioned primary energy. The supply cost was defined as the cost at the fueling station and included the aforementioned energy cost and the plant cost described in Section 2.4.

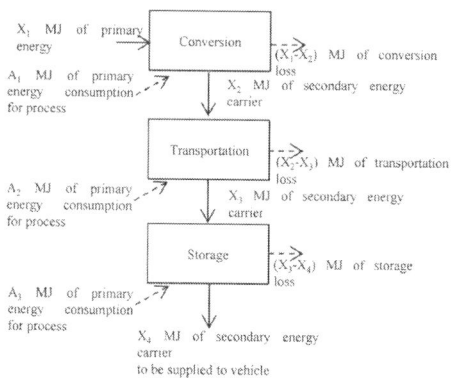

Figure 1: Energy balance for each energy supply path.

Timeframe

The evaluation was conducted for year 2010 and 2020. The 2010 evaluation determined the relative merits of the AESs using current costs and technological performance. The 2020 evaluation investigated how future technological advancements and cost reduction could impact the three evaluation indicators.

Data Set

Capital Cost and Operation and Maintenance (O&M) Cost

In this study, the fixed component of the supply cost, namely the capital cost and the O&M cost, was estimated using the calculation rule shown in Table 2[22]. The initial capital cost was calculated as the sum of (a) the plant and equipment cost, and (b) the cost of general facilities, engineering, permitting and start-up, contingencies, working capital, land and miscellaneous expense, all of which were expressed as a percentage of the plant and equipment cost. The initial capital cost was then distributed as a levelized annual burden equal to 13.4%[1] of the total initial capital cost. The annual O&M cost represented 6% of the initial capital cost.

Table 2: Conditions for estimating supply cost (excluding energy cost)

Item	Condition
(a) Initial capital cost calculation	
General facilities	20% of equipment cost
Engineering, permitting and start-up	10% of equipment cost
Contingencies	10% of equipment cost
Working capital, land and misc.	7% of equipment cost
(b) Annual capital cost calculation	
Annual capital charge	13.4% of initial capital cost

(c) Annual operation cost calculation	
Fixed O&M cost	6% of initial capital cost

Table 3 shows the assumptions for the plant and equipment cost in 2010. The plant cost must be carefully defined because costs for the same type and capacity fluctuate from one year to another due to the world's changing economic circumstance and because the unit capital cost usually decreases as the total capacity increases. In order to avoid the ambiguities introduced by various literature for this cost, a unified approach for cost estimation was employed by using the following calibration. First, cost data obtained from the literature was calibrated to the figure in year 2010 using the Chemical Engineering Plant Cost Index [23]. Second, the capacity of the energy conversion plants was standardized to 1161 million MJ/h for the off-site energy conversion plants (which was equivalent to a large scale ammonia production plant with a capacity of 1,500 t/d), and 3,239 MJ/h for the on-site plants (which was equivalent to the typical capacity of a natural gas reformer installed in a hydrogen fueling station in Japan).

Table 3: Assumptions for plant and equipment costs

Plant/equipment	Capacity	Capacity (In MJ)	Cost for 2010 (Million JPY)
Ammonia production plant (Haber–Bosch Process)	1,500 t/d	1,161,063 MJ/h	6,361
Electrochemical ammonia synthesis equipment	174 kg/h	3,239 MJ/h	659
High pressure ammonia truck	0.3t	6,186 MJ	4.4
High pressure ammonia storage tank	1t	18,577 MJ	4.5
Hydrogen production facility at refinery	2,580,858 Nm³/d	1,161,063 MJ/h	4,217
Methane reforming equipment	300 Nm³/h	3239 MJ/h	211

Electrolysis equipment	300 Nm³/h	3,239 MJ/h	200
High pressure hydrogen truck	2,740 Nm³	29,584 MJ	30
High pressure hydrogen storage tank	1,536 Nm³	16,584 MJ	44
Rapid electric charger	50 kW	50 kW	2.4

Table 4 shows the assumptions that were made for the cost reduction target for the year 2020, using a publication of the New Energy and Industry Development Organization (NEDO), an affiliate of the Japanese government [4].

Table 4: Assumption of equipment cost reduction for 2020

Equipment	Cost reduction rate (%)
Hydrogen production equipment	50
PSA	30
Compressor	30
Tank	20
Dispenser	50

The annual exchange rate of 87.81 JPY/USD in 2010 was used throughout the study [24].

Energy Efficiency

Table 5 sets out the current and future energy efficiencies of the technologies considered in this study [4],[5], [15], [19], [25] and [26]. The energy efficiency expressed as how much energy is consumed to produce, transport or store unit amount of energy carrier is an important technical data, and its future change will induce a significant change in AES's comparative advantage. It must be noted that no efficiency improvement was assumed for several mature technologies including transportation, storage, hydrogen production from refinery, natural gas reforming, and electric charger, although some future cost reduction was projected, as discussed in Section 2.4.1.

Table 5: Assumptions on energy efficiency for 2010 and 2020

Technology	Item	2010	2020
Ammonia production plant (Haber–Bosch process)	CH_4 feedstock required for t-NH_3	25,100 MJ	20,247 MJ
	CH_4 fuel required for t-NH_3	8,200 MJ	6,615 MJ
Electrochemical ammonia synthesis	Electricity feedstock required for kg-NH_3	11.83 kWh (2.5 V)	7.10 kWh (1.5 V)
	Electricity required for kg-N_2 using ASU[a]	0.68 kWh	0.68 kWh
	Electricity required for kg steam generation	0.76 kWh	0 kWh
By-produced H_2 at refinery. By-produced H_2 at refinery	Naphtha consumption for kg-H_2	2.61 kg	2.61 kg
	Oil fuel required for kg-H_2	2.17 kg	2.17 kg
	Electricity required for kg-H_2	0.47 kWh	0.47 kWh
Methane reforming	CH_4 feedstock required for kg-H_2	3.67 kg	3.67 kg
	Electricity required for kg-H_2	6.53 kWh	6.53 kWh
Electrolytic hydrogen production	Electricity consumption for kg-H_2	57.84 kWh	43.38 kWh
Ammonia truck	Gasoline fuel efficiency	4.04 km/L	4.04 km/L
Hydrogen truck	Gasoline fuel efficiency	4.04 km/L	4.04 km/L
Ammonia storage tank	Electricity required for operation	20 kJ/kg-NH_3	20 kJ/kg-NH_3
Hydrogen storage tank	Electricity required for operation	1.08 kWh/kg-H_2	1.08 kWh/kg-H_2

[a]ASU: air separation unit.

Operational Conditions

It was assumed that plants operate 13 h/day and 365 days/year in accordance with the literature [26]. The transportation distance for the off-site energy supply pathways was uniformly set at 50 km. This distance was based the typical geography in Japan, where most petrochemical and chemical plant complexes are situated approximately 50 km away from urban area with high energy demands. The energy carrier was stored principally for a day before being used to fulfill the demand. Storage for more than a day was also considered in Section 3.3 in order to examine energy systems that are resilient to supply disruptions.

Energy Price and Unit CO_2 Emissions

Table 6 lists the fossil fuel prices and CO_2 emissions used in this study. The gas price in 2010 corresponded to the average retail price for non-residential customers, and the naphtha and gasoline prices in 2010 corresponded to average wholesale prices [27]. The future prices in 2020 were calculated multiplying current price with the projected percentage increase in the imported energy price published by the Japanese government [28]. The figures given by the Ministry of Environment were used for the unit CO_2 emissions [29].

Table 6: Unit price and CO_2 emissions of fossil fuels

Type	Unit price (JPY/1000 kcal)		Unit CO_2 emissions
	2010	2020	2010/2020
CH_4	5.57	6.51	2.22 kg-CO_2/Nm^3
Naphtha	6.23	8.49	2.24 kg-CO_2/L
Gasoline	14.63	19.93	2.32 kg-CO_2/L

The electricity price and the CO_2 emissions per kWh were provided in Table 7, which were based on the long-term energy strategy scenario under debate in the Japanese government. One must bear in mind that prospects for the future prices and unit CO_2 emissions are unclear because of the significant uncertainty in the long-term nuclear energy policy. Table 7 indicates that both the electricity price and the unit CO_2 emissions depend on the options associated with future electricity generation mix including nuclear power generation [28]. This study mainly relied on the intermediate scenario, Case 1. This scenario presumed that no new nuclear power plant would be built and that existing nuclear power stations would be allowed to operate until the end of their legally authorized economic life.

Table 7: Assumptions on CO_2 emissions and electricity price for several electricity generation mix scenarios

Case		2010	2020			
		Case 1	**Case 2-1**	**Case 2-2**	**Case 3**	
Electricity mix	Nuclear	26.4%	20.9%	0.0%	13.9%	23.5%
	Coal	24.5%	23.2%	27.4%	24.7%	22.3%
	LNG	29.1%	29.7%	39.9%	33.2%	28.1%
	Oil	9.5%	7.7%	12.7%	8.3%	7.5%
	Renewables	10.5%	18.5%	20.0%	20.0%	18.5%
CO_2 emissions (kg-CO_2/kWh)		0.423	0.402	0.523	0.436	0.386
Electricity price (JPY/kWh)		13.65	16.4	16.9	16.9	16.4

The unit electricity price in 2010 was set as the actual average price for non-residential customer in the same year [30]. The electricity price in 2020 is estimated to be 2.8 JPY/kWh higher than the 2010 price, assuming the expected increase in fuel price in Case 1 is fully reflected to the generation cost and then to the end user price [28].

The relative merits of AESs versus HESs and EESs were further analyzed using a sensitivity analysis for the unit CO_2 emissions to evaluate the potential impact of change in the power generation mix from Case 1 to Case 3.

RESULTS AND DISCUSSION

Analysis for the Year 2010

Fig. 2 illustrates the results of the three indicators calculated with respect to the feedstock, the conversion, the transportation, and the storage process, respectively. It should be noted that there was no transportation cost for the on-site pathways because these pathways did not involve transportation. Also, it must be borne in mind that EES assumes the use of grid electricity which involves all the functions from production to distribution; therefore, the indicators were expressed as an integrated value including all the feedstock, the conversion and the transportation.

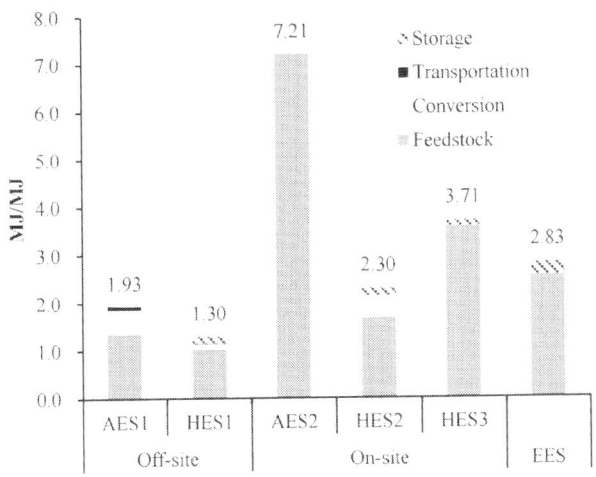

(a)

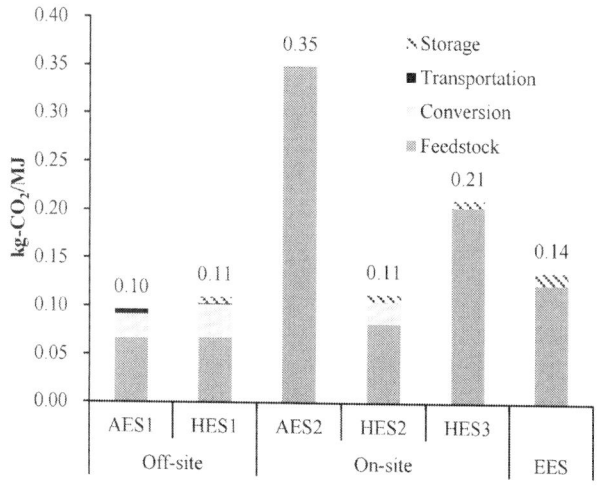

(b)

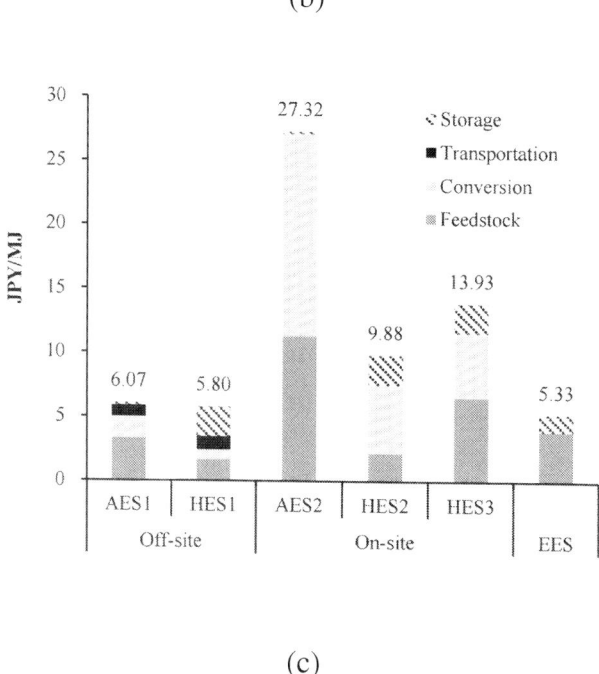

(c)

Figure 2: Primary energy consumption, CO_2 emission and cost per MJ secondary energy supply for 2010.

As to the primary energy consumption per energy carrier supplied, it was clearly observed that the energy consumption for converting primary energy to secondary energy carrier was the determinant of the efficiency (the lower the value, the higher the efficiency of the entire energy pathway). The EES exhibited the lowest efficiency among the off-site pathways. This is because whereas hydrogen or ammonia can be produced from fossil fuels at a relatively high conversion efficiency (of approximately 90%) in a large-scale chemical plant, the conversion efficiency from primary energy to electricity is as low as 0.41 (the actual figure in 2009 in Japan). For the on-site pathways, the efficiencies of AES2 and HES3 were apparently lower due to the energy loss in the duplicated conversion processes, i.e., power generation and ammonia or hydrogen synthesis using electrolysis.

A similar trend was observed for the CO_2 emissions per energy carrier supplied: the CO_2 emitted in the conversion process determined the relative merit of each pathway. For the off-site pathways, the EES exhibited a larger value because of the aforementioned nature of lower conversion efficiency from primary energy to electricity. Regarding the on-site pathways, the emissions for AES2 and HES3 were obviously higher due to the energy losses in the duplicated conversion processes.

A cost analysis showed that EES was the most cost competitive pathway, indicating the exertion of an economy of scale for the existing electricity infrastructure. HES1 showed a natural advantage over AES1 because hydrogen production currently only involves reforming process, whereas ammonia production inevitably includes NH_3 synthesis process using Haber–Bosch technique other than reforming. However, this advantage is compromised because the ammonia pathway uses fully commercialized and cost competitive transportation and storage, which has not been achieved for the hydrogen pathways thus far. The on-site analysis showed that the pathways involving electrolysis were less competitive. This result is attributed to several factors including the lower efficiency of electrolysis, which leads to higher electricity consumption, and the current high cost of electrolytic equipment. Given the current technical and cost competitiveness of the respective plant and equipment, the AES does not exhibit apparent advantages over the HES or the EES for short-term storage periods such as 1 day. The most commercialized current ammonia production technology comprises the two main energy consuming steps, fossil fuel reforming to hydrogen, and NH_3 synthesis by the Haber–Bosch process at a high temperature and high pressure. This implies inevitable disadvantage with efficiency when comparing with hydrogen production, which only involves fossil fuel reforming to hydrogen. The electricity-based pathways have already been formulated as a reliable infrastructure with an economy of scale.

Analysis for the Year 2020

The impacts of foreseeable improvement in economic and technical performance on the three indicators have been analyzed as shown in Fig. 3. Note that no economic or technical improvement was assumed to the EES because of the maturity of the existing systems. Also no improvement for CO_2 emissions was observed to the EES because the fossil fuel based power generation will be more dominant in 2020 due to the suspension of nuclear power expansion caused by the catastrophic nuclear accident in 2011. A significant improvement was observed for the AES and the HES for most of the indicators because of significant progress in both technological performance and cost reduction. In particular, pathways utilizing electrolysis such as the AES1 and the HES3 exhibited a remarkable improvement in all the indicators following the expected advancement of electrolysis technology.

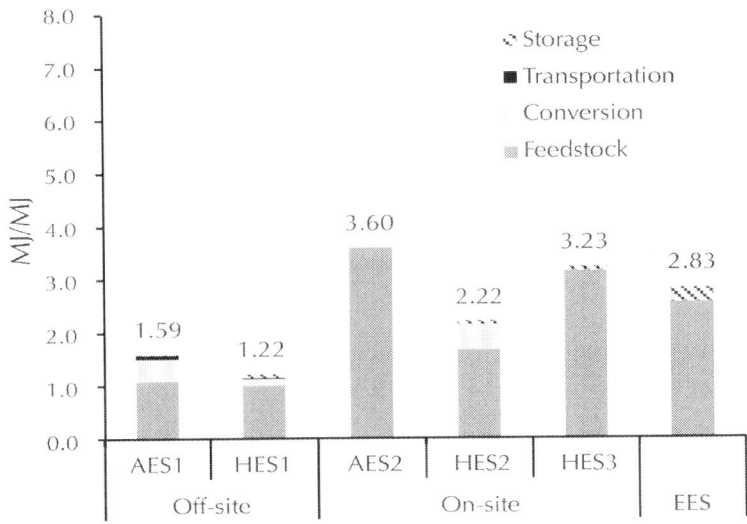

(a)

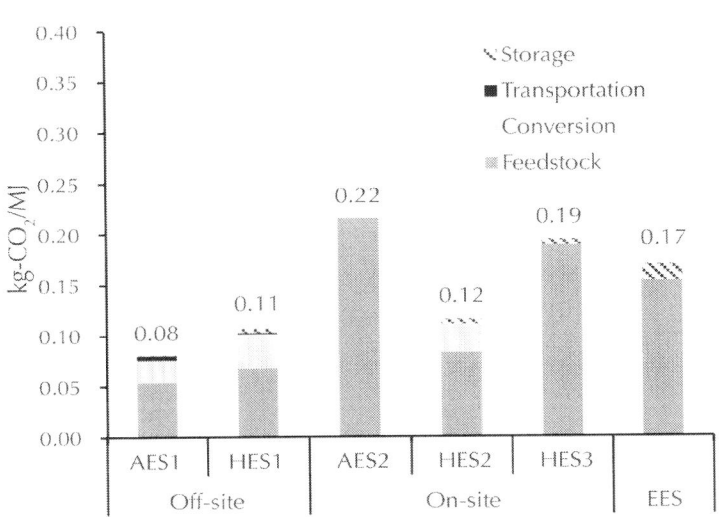

(b)

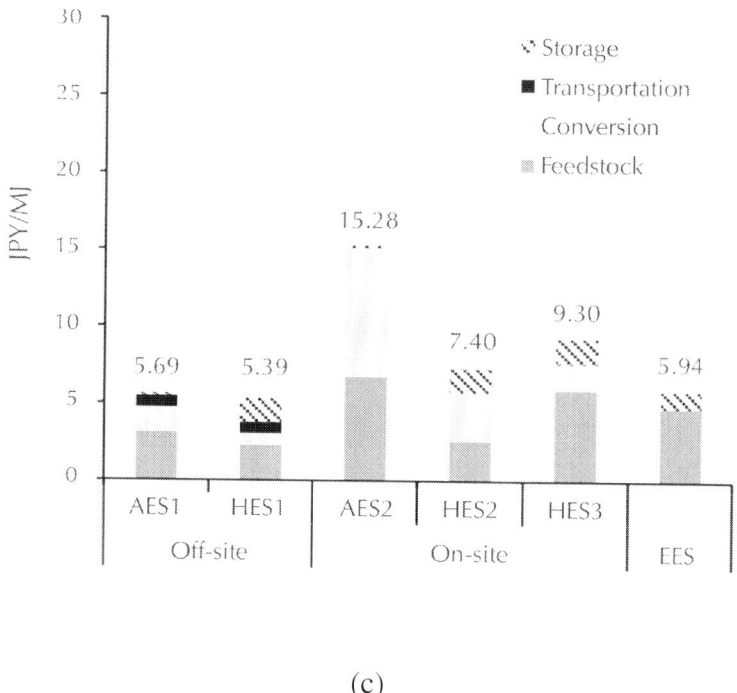

(c)

Figure 3: Primary energy consumption, CO_2 emission and cost per MJ secondary energy supply for 2020.

Nevertheless, even after considering the aforementioned future improvements, there was little change in the comparative merits of the AES over the HES and the EES. In particular, the AES was inferior to the EES in terms of the supply cost, which is a determining factor in one's technology selection. In addition to the indicator analysis, it should be mentioned that the supply for the relatively competitive AES1 is technically constrained by the redundant capacity in domestic ammonia production, as described in Section 2.1.

Impact to the supply cost due to the change in the equipment cost for the year 2020 by ±10% was analyzed. The range of the cost was calculated to be, in terms of JPY/MJ, 5.69 ± 0.10 for AES1, 5.39 ± 0.24 for HES1, 15.28 ± 0.85 for AES2, 7.40 ± 0.37 for HES1, 9.3 ± 0.32 for HES3, and 5.94 ± 0.06 for EES. This result indicates the difference by ±10% does not influence most of the cost order of each path, and thus does not affect the above discussion on the cost competitiveness.

Also, impact to the primary energy consumption per energy carrier due to the change in the efficiency improvement in Table 5 by ±10% was analyzed. The range of the consumption was calculated to be, in terms of MJ/MJ, 1.59 ± 0.04 for AES1, 3.60 ± 0.27 for AES2, and 3.23 ± 0.10 for HES2, while remained unchanged for the other supply paths considering the technology has been already matured. This result indicates the difference by ±10% does not affect the above discussion on energy efficiency competitiveness.

Analysis for Long-term Storage

The threats to the continuous energy supply by the great east Japan earthquake and the subsequent malfunctioning of the electricity infrastructure highlighted Japan's energy vulnerability that appeared as an energy supply disruption of more than several days. An energy system that is resilient to external shocks requires a technically mature and economically affordable energy storage system to ensure energy supply to final customers during supply disruptions of not only minutes or hours but also days. To evaluate the capability of the AES as a long-term storage medium, the supply cost for various numbers of storage days was investigated. The result is shown in Fig. 4. The significant increase in the supply cost was attributed to the longer carrier detention time, which affected recurrent use of the storage facility and therefore caused a spike in the fixed cost per supplied energy carrier. For the off-site pathways, the cost competitiveness of AES became apparent for longer storage periods. For the on-site pathways, the AES became more competitive than the HES for more than 10 storage days, and even became almost competitive with the EES at 20 storage days.

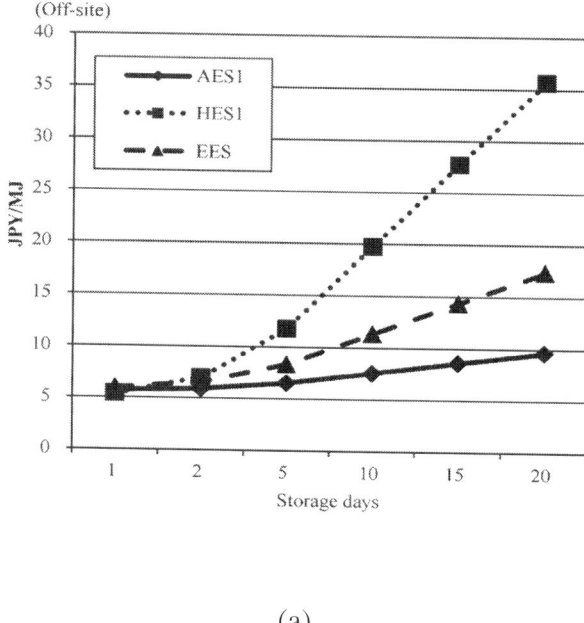

(a)

(b)

Figure 4: Relationship between the number of storage days and energy supply cost per MJ for 2020.

Sensitivity Analysis for Co_2 Emissions Intensity of Electricity Generated

Fig. 5 illustrates how the CO_2 indicator for each energy supply path was influenced by the scenarios with different unit CO_2 emissions of electricity. The main observation was that the paths that relied heavily on electricity such as the AES2, the HES2, and the EES were more sensitive than those that relied on fossil fuels such as the AES1, the HES1 and the HES2. This indicates that the energy supply paths with a lower reliance on electricity more accurately describe future CO_2 emissions in Japan, which is amidst increasing energy uncertainties, and would thus be helpful in the country formulating a CO_2 abatement strategy that is resilient to dynamic change in the electricity generation mix.

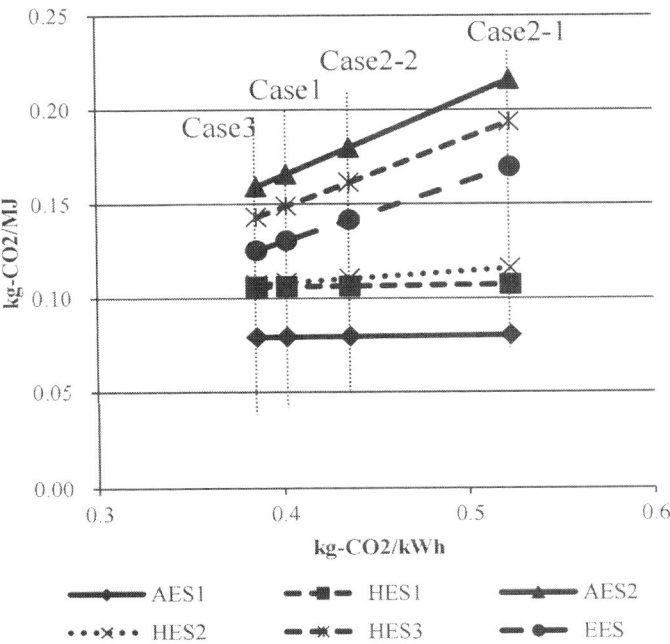

Figure 5: Sensitivity analysis of unit CO_2 emissions per MJ secondary energy supply to CO_2 emissions intensity per kWh of electricity.

Advantage and Challenges of Ammonia Energy Systems

In light of the aforementioned analysis, the relative advantages and challenges associated with the AES are discussed below.

Advantages

First, the potential superiority of the AES can be attributed to the technical and commercial maturity of ammonia transportation and storage. In the ammonia industry, ammonia is commonly produced in a large-scale plant from natural gas or coal, after which it is transported through pipeline or by a truck with a high pressure vessel, stored in different vessels, and delivered to the final customers, namely large-scale industrial consumers such as fertilizer companies. Thus, as illustrated in Fig. 4, the AES exhibits a comparative advantage over the HES and the EES when long-term storage is needed in the energy system. That is, ammonia should be used as an energy carrier to improve the overall security of the energy supply, by making the system resilient to energy supply disruption through a convenient and inexpensive means of energy storage.

Second, the AES has the potential to significantly abate CO_2 emissions. The AES2 uses an electrolytic technique to synthesize ammonia from water and nitrogen in the air, and does not consume any fossil fuels. Thus, using completely carbon free primary energy sources would minimize the CO_2 emissions from the AES2 [6]. There are some other carbon free technologies including catalytic reduction of nitrogen to ammonia in the presence of a molybdenum complex [7]. Although fundamental research and development are still required, these novel techniques may represent a means of developing an ammonia energy system with minimum or even no CO_2 emission.

Challenges

As previously discussed, the clear superiority of the AES is limited to energy systems requiring a sizable energy storage for energy security purposes. To achieve technical and economic competitiveness for small storage requirements (e.g., 1 day), both the AES1 and the HES2 need to fill in the gap with regard to all the three indicators.

A key issue is to develop demand side technologies that utilize ammonia, including the ammonia combustion engine and ammonia-based fuel cells. Fig. 6 illustrates the amount of primary energy needed to meet a transport service demand of 100 km, based on the same AES/HES/EES pathways, and assuming fuel efficiencies of 175 MJ/km for an ammonia ICE vehicle, 85.2 MJ/km for a hydrogen fuel cell vehicle, and 44.6 MJ/km for an electric vehicle [31]. The AES evidently required more primary energy than the HES or the EES to fill the same service demand. Currently the fuel efficiency of hydrogen or electricity fueled vehicles is drastically higher than that of vehicles fueled by ammonia.

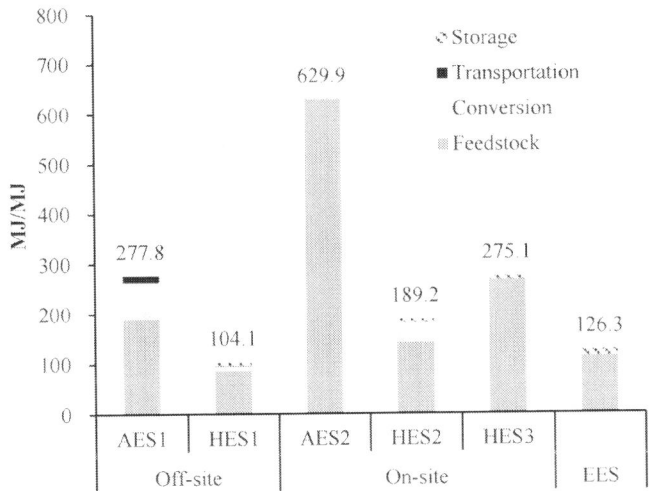

(a)

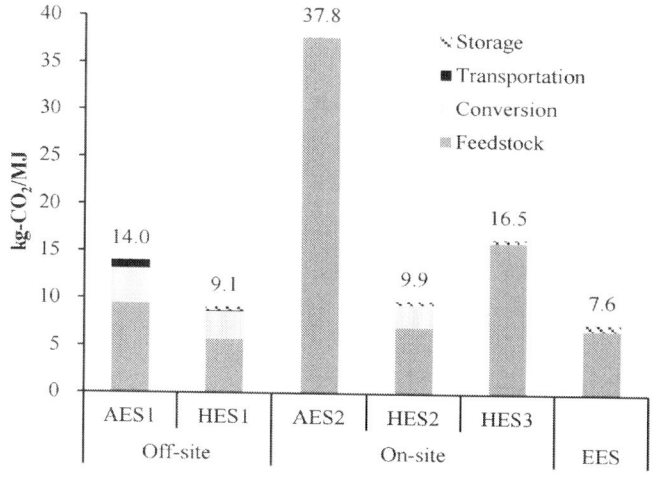

(b)

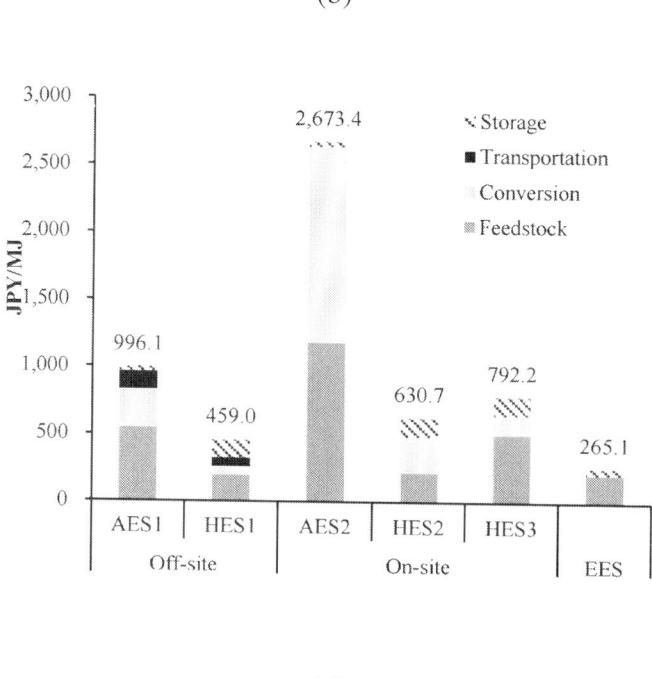

(c)

Figure 6: Primary energy consumption, CO_2 emission and cost per 100 km vehicle service demand for 2020.

Another challenge is posed by the social acceptance of ammonia energy systems by the public, on which, as far as the authors are concerned, no study has been conducted thus far. Considering that ammonia is a weakly toxic substance, measures must be formulated for potential ammonia leakage to secure public acceptance and support or the AES [21].

R&D

Addressing the challenges posed by ammonia energy systems requires extensive R&D activities. First, R&D should focus on technological and economic improvements in the ammonia production process which consists of most of the ammonia energy systems rather than the already matured transportation and storage processes. For instance, the efficiency improvements of the Haber–Bosch process need to be pursued by exploring utilization of the process waste heat. Although there is an industry prospect that the primary energy required to produce ammonia will be reduced from 33.1 GJ/t-NH$_3$ in 2002 to 26.7 GJ/t-NH$_3$ in 2014, our analysis indicated that a further cost reduction of both capital cost and fuel cost by 23%, respectively, would be required for the AES to be economically competitive with the HES and the EES for the 1-day storage case [5]. Also, to foresee substantial CO$_2$ reduction by adopting the AES, new carbon free NH$_3$ production techniques, such as the direct synthesis of ammonia from water and nitrogen, should be researched further to evaluate whether such processes are technically viable for future commercialization.

Second, an attention should be paid to the R&D aiming to address current immature demand side technologies, namely ammonia-driven fuel cell vehicles and ammonia ICE vehicles. Key R&D issues include the improvement of ammonia combustion characteristics by hydrogen addition for ICE vehicles[32] and contaminant removal in PEFC membranes and catalyst layers with NH$_3$[33].

Third, the social acceptance of ammonia energy systems should be studied to effectively formulate safety measures for potential ammonia leakage.

CONCLUSIONS

This study assessed the relative merits of ammonia energy systems for vehicle use in Japan by estimating the three indicators (the primary energy consumption per energy carrier supplied, the CO_2 emissions per energy carrier supplied and the supply cost per energy carrier supplied). The potential benefits and challenges associated with using ammonia as an energy carrier were discussed, and an R&D agenda was presented to improve ammonia's relative competitiveness over hydrogen and electricity.

Using ammonia as an energy carrier was demonstrated to be competitive in terms of efficiency, CO_2 emissions and supply cost for energy systems requiring fairly large numbers of storage days. Thus utilizing ammonia in part of an energy system should enhance continuity of the energy supply in a country or region facing the insecurity of supply.

In the meantime, further technical improvements and cost reduction associated with both conventional and unconventional ammonia production were found to be imperative for using ammonia in a normal energy system. The efficiency improvement of Haber–Bosch process needs to be pursued by exploring better utilization of process waste heat. New ammonia production techniques with the potential to be CO_2-free, including direct ammonia synthesis from water and nitrogen, should be researched further, and these technologies should be validated in terms of future commercialization.

In this study, the consideration was limited to the adaptation of ammonia energy systems to transportation demand. Also emphasis was made on a bottom-up economic and environmental assessment, such that the evaluation of the security of supply could be improved. These issues will be addressed in our future work by establishing an energy model enabling a more detailed quantitative evaluation on not only economic and environmental aspects, but also the security of the energy supply for ammonia energy systems.

ACKNOWLEDGMENTS

The authors would like to sincerely thank Dr. Yasuhiko Ito, Professor Emeritus at Kyoto University for his valuable comments and suggestions. The authors also would like to thank Mr. Masaru Adachi for his analysis of the ammonia energy system.

REFERENCES

1.   Onoue K, Murakami Y, Sofronis P. Japan's energy supply: mid-to-long-term scenario e a proposal for a new energy supply system in the aftermath of the March 11 earthquake. Int J Hydrogen Energy 2012;37:8123e32.

2.   Vivoda V. Japan's energy security predicament post-Fukushima. Energy Policy 2012;46:135e43.

3.   New Energy and Industrial Technology Development Organization. Battery roadmap 2010. Available from: http://www.nedo.go.jp/content/100153876. pdf; 2010

4.   New Energy and Industrial Technology Development Organization. NEDO fuel cell and hydrogen technology roadmap 2010. Available from: http://www. nedo.go.jp/news/other/FF_00059.html; 2010

5.   The Canadian Fertilizer Institute, The Canadian Industry Program for Energy Conservation. Benchmarking energy efficiency and carbon dioxide emissions. Ottawa: The Canadian Industry Program for Energy Conservation; 2007.

6.   Murakami T, Nohira T, Goto T, Ogata YH, Ito Y. Electrolytic ammonia synthesis from water and nitrogen gas in

molten salt under atmospheric pressure. Electrochim Acta 2005;50:5423e6.

7.   Arashiba K, Miyake Y, Nishibayashi Y. A molybdenum complex bearing PNPtype pincer ligands leads to the catalytic reduction of dinitrogen into ammonia. Nat Chem 2011;3:120e5.

8.   Klerke A, Christensen CH, Norskovb JK, Vegge T. Ammonia for hydrogen storage: challenges and opportunities. J Mater Chem 2008;18:2304e10.

9.   Tsubota M, Hino S, Fujii H, Oomatsu C, Yamane M, Ichikawa T, et al. Reaction between magnesium ammine complex compound and lithium hydride. Int J Hydrogen Energy 2010;35:2058e62.

10.   Veltman M. Developing fuel injection strategies for using ammonia in direct injection diesel engines. Iowa State Univeristy; 2009.

11.   Saika T. Burning velocities of ammoniaehydrogen fuel. In: Proceedings of the fifth annual meeting of Japan Institute of Energy, 5; 1996. pp. 235e7.

12.   Yang J, Muroyama H, Matsui T, Eguchi K. Development of a direct ammoniafueled molten hydroxide fuel cell. J Power Sources 2014;245:277e82.

13.   Suzuki S, Muroyama H, Matsui T, Eguchi K. Fundamental studies on direct ammonia fuel cell employing anion exchange membrane. J Power Sources 2012;208:257e62.

14.   Ma Q, Peng R, Lin Y, Gao J, Meng G. A high-performance ammonia-fueled solid oxide fuel cell I. J Power Sources 2006;161:95e8.

15.   Bartels JR, Pate MB. A feasibility study of implementing an ammonia economy. Iowa State University; 2008.

16.   Zamfirescu C, Dincer I. Ammonia as a green fuel and hydrogen source for vehicular applications. Fuel Process Technol 2009;90:729e37.

17.   The Department of Energy. Potential roles of ammonia in a hydrogen economy, a study of issues related to the use

of ammonia for on-board vehicular hydrogen storage. Washington D.C.: U.S. Department of Energy; 2006.

18.   T-Raissi A. Technoeconomic analysis of area II hydrogen production e part II. Hydrogen from ammonia and ammoniaeborane complex for fuel cell applications. In: Proceedings of the 2002 U.S. DOE hydrogen program review; 2002. NREL/C P-610e32405.

19.   Japan Fertilizer and Ammonia Producers Association. Ammonia supply and demand statistics. Available from: http://www.jaf.gr.jp/; 2014

20.   Ministry of Land Infrastructure and Transportation. Fuel efficiency standards for vehicles. Available from: http://www.mlit.go.jp/common/000037091.pdf; 2011

21.   Appl M. Ammonia: principles and industrial practice. Weinheim. New York: Wiley-VCH; 1999.

22.   National Research Council (U.S.). Committee on Alternatives and Strategies for Future Hydrogen Production and Use, National Academy of Engineering, National Academy of Sciences (U.S.). The hydrogen economy: opportunities, costs, barriers, and R&D needs. Washington, D.C: National Academies Press; 2004.

23.   Access Intelligence LLC. Chemical engineering plant cost index. Available from: http://www.che.com/pci/; 2011

24.   Mitsubishi UFJ Research & Consulting. Exchange quotations-yearly averages. Available from: http://www.murc.jp/fx/yearend/index.php?id¼2010; 2011

25.   New Energy and Industrial Technology Development Organization. Electrolytic ammonia synthesis from water and nitrogen in molten salts under atmospheric pressure and the ammonia energy system. Japan: New Energy and Industrial Technology Development Organization; 2011. Report No.: 09008065-0.

26.   Japan Hydrogen and Fuel Cell Demonstration Project. FCV-oriented hydrogen infrastructures demonstration study (stage

II) final report. Available from: http://www.jari.or.jp/jhfc/data/report/pdf/tuuki_phase2_01.pdf; 2011

27. The Institute of Energy Economics Japan. EDMC handbook of energy and economic statistics in Japan 2012. Tokyo: The Energy Conservation Center, Japan; 2012.

28. The Energy and Environment Council. Options for energy and the environment, the energy and environment council decision on June 29, 2012. National Policy Unit, Cabinet Secretarist; 2012.

29. Ministry of Environment. Unit GHG emissions for calculation, reporting and disclosure. Available from: http://ghg-santeikohyo.env.go.jp/files/calc/itiran. pdf; 2010

30. Agency for Natural Resources and Energy. On competition in liberalized retail electricity market. Available from: http://www.meti.go.jp/committee/   sougouenergy/sougou/denryoku_system_kaikaku/002_s01_01_18.pdf; 2012

31.   Dincer I, Rosen MA, Zamfirescu C. Economic and environmental comparison of conventional and alternative vehicle options. In: Electric and hybrid vehicles, power sources, models, sustainability, infrastructure and the market. Oxford: Elsevier; 2010. pp. 1e17.

32. Saika T, Nakamura M, Nohara T, Ishimatsu S. Study of hydrogen supply system with ammonia fuel. JSME Int J B Fluid Therm Eng 2006;49:78e83.

33. Zhang XY, Pasaogullari U, Molter T. Influence of ammonia on membranee electrode assemblies in polymer electrolyte fuel cells. Int J Hydrogen Energy 2009;34:9188e94.

Citations

CHAPTER 1

Sukhvinder P. S. Badwal, Sarbjit S. Giddey, Christopher Munnings, Anand I. Bhatt and Anthony F. Hollenkamp doi: Emerging electrochemical energy conversion and storage technologies 10.3389/fchem.2014.00079.

CHAPTER 2

Mohammed Dwidar, Jae-Yeon Park, Robert J. Mitchell, and Byoung-In Sang, "The Future of Butyric Acid in Industry," The Scientific

World Journal, vol. 2012, Article ID 471417, 10 pages, 2012. doi:10.1100/2012/471417.

CHAPTER 3

J. Burger and M. Gochfeld, "Knowledge and Perceptions of Energy Alternatives, Carbon and Spatial Footprints, and Future Energy Preferences within a University Community in Northeastern US," Energy and Power Engineering, Vol. 5 No. 4, 2013, pp. 322-331. doi: 10.4236/epe.2013.54033.

CHAPTER 4

Chang, S. (2014) Anaerobic Membrane Bioreactors (AnMBR) for Wastewater Treatment. Advances in Chemical Engineering and Science, 4, 56-61. doi: 10.4236/aces.2014.41008.

CHAPTER 5

Yasunori Kikuchi, Seiichiro Kimura, Yoshitaka Okamoto, Michihisa Koyama, A scenario analysis of future energy systems based on an energy flow model represented as functionals of technology options, Applied Energy, Volume 132, 1 November 2014, Pages 586-601, ISSN 0306-2619, http://dx.doi.org/10.1016/j.apenergy.2014.07.005.

CHAPTER 6

Daisuke Miura, Tetsuo Tezuka, A comparative study of ammonia energy systems as a future energy carrier, with particular reference to vehicle use in Japan, Energy, Volume 68, 15 April 2014, Pages 428-436, ISSN 0360-5442, http://dx.doi.org/10.1016/j.energy.2014.02.108.

Index